T0133942

MECHANICS

MECHANICS

Ashok S. Pandit

CRC Press
Boca Raton   London   New York   Washington, D.C.

Narosa Publishing House
New Delhi   Chennai   Mumbai   Calcutta

Ashok S. Pandit
Department of Mathematics
University of Mumbai
Mumbai-400 098, India

Library of Congress Cataloging-in-Publication Data:

A catalog record for this book is available from the Library of Congress.

This book contains information obtained from authentic and highly regarded sources.
Reprinted material is quoted with permission, and sources are indicated. Reasonable
efforts have been made to publish reliable data and information, but the author and the
publisher cannot assume responsibility for the validity of all materials or for the
consequences of their use.

Exclusive distribution in North America only by CRC Press LLC

Direct all inquiries to CRC Press LLC, 2000 N.W. Corporate Blvd., Boca Raton,
Florida 33431. E-mail: orders@crcpress.com

Copyright © 2001 Narosa Publishing House, New Delhi-110 017, India

No claim to original U.S. Government works
International Standard Book Number 0-8493-0979-4
Printed in India.

Preface

It is said that Archimedes could hold at bay an invading Roman fleet in Syracus harbour for three years with devastating catapults and big-beaked ship-biting iron claws of his own design. The Roman general Marcellus, after his first repulse, called Archimedes "the geometric Briareaus (a mythological monster with hundred arms) who uses our ships like cups to ladle water from the sea !"

I think, the same description, the geometric Briareus, is applicable to the subject of discourse of my text-book: Classical Mechanics. Indeed, being the Newtonian model of the entire physical world, it is very extensive and powerful like the mythological monster. Secondly, be it in its primitive Newtonian version which makes use of the Euclidean geometry of vectors or the present day symplectic geometric avatar, it is manifestly geometric in nature. It draws its strength from geometry and in turn, stimulates not only geometry but entire mathematics. Consequently, a course in classical mechanics is a course in two disciplenes namely, mathematics and physics. A mathematically sound treatment of mechanics provides a student a deeper grasp of all of physics. On the other hand, many of the geometric concepts are better understood when visualised in terms of dynamical notions such as the trajectory of a particle, its velocity, its acceleration and so on.

There is another important aspect involved in the study of classical mechanics which is explained below (though very briefly).

It is known from physiology that human brain is compartmentalized into two hemi-spheres. These spheres perform different but complementry functions. The right hemi-spliere is superior at dealing with the geometrical aspects of any material phenomenonical concepts. The left hemi-sphere provides the logical/ analytical steps from one point to another in the thinking process. The right hemi-sphere gives geometric vision needed to see the final goal and the landmarks on the way. A study of classical mechanics is a marvellous means to develop and integrate both these geometric and logical/ analytical powers of the brain.

It is for such reasons that all universities reserve a place for classical mechanics at the undergraduate level in all the three disciplines, namely, mathematics, physics and engineering.

This book is an introduction (at the undergraduate level) to the fascinating subject of classical mechanics. The treatment of the subject is mathematical, but the mathematics used is elementary. In fact, all the mathematics we use is basic vector algebra and a little bit of differential calculus of vector valued functions of a real variable. We have avoided explaining the underlying geometry in an explicit manner. Instead, all the geometric discussion appears in a kinematical disguise. I belive that such a treatment is benificial not only to mathematics students, it is most straight-forward approach to the subject from the point of view of physics/ engineering students as well.

Each chapter begins with an introduction to the concepts involved in the topic of the chapter. This is followed by precise definitions, propositions and the theorems covering the theme of the chapter. The results are further elucidated by illustrative examples and solved problems. There are quiet a few exercises at the end of each chapter.

The material of this book is based on a lecture course which I gave a few years back to undergraduate students of mathematics and physics studying in Mumbai colleges. A set of lecture nots was prepared for the course. NBHM gave financial help for the preparation of the lecture notes. I take this opportunity to express my gratitude to NBHM authorities. I had several opportunities during the last few years to give short courses and problem sessions under the auspices of Mathematics Department of Mumbai University and Mumbai Mathematics Colloquium. I thank both of them.

A number of people helped me during the preparation of the book: Professor S.S. Sane and Professor R.C.Cowsik allowed me to use computer and laser printing facilities at the department. On several occasions, Professor Cowsik had to come to my rescue when the typing work developed snags. Miss Sunita Shringarpure has done the onerous task of typing (and retyping, several times!) the text. Without her assistance, this book would have never come out. Shri Nandesh of INTEGRA (Media Graphics) has drawn the diagrams. I thank all these people.

ASHOK S. PANDIT

Contents

Chapter 1

Basic Concepts of Motion

There is a story about how a certain mathematician would begin his course in logic "Logic is the science of laws of thought" he would declaim. "Now I must tell you what science is, what law is and what thought is. But I will not explain what 'of' means".

Yu. I. Manin

1.1 Introduction

Mechanics is the branch of science which deals with the motion of material objects.

Let V denote the physical space.

Thus, V is the immense, endless emptiness in which all material objects are found dwelling and moving about.

Let us consider the motion of a single material object. (From now-onwards, we will refer to it as an *object* instead of the longish term: material object.) At any instant, the object occupies a definite position in V; it is its *instantaneous position*. When a change occurs in a continuous manner in its positions as the time progresses, we say that the *object is in motion*.

Thus the concept of motion inter-relates space, time and the matter.

Let I denote an interval of time.

We consider the motion of the object during the time interval I. Clearly, the motion is nothing but the chronological sequence describing where the object was located in V at each instant $t \in I$. In other words, the motion is the *parameterization* of position of the object by the corresponding instants drawn from I.

Being thus a time dependent process, the motion of the object gives rise to some more time dependent ingradiants such as *velocity, acceleration* and so on.

1

At this stage, let us introduce a new dynamical term. We consider (an ordered) pair consisting of the position and the velocity of the object at an instant. We call the pair the instantaneous *state of motion* of the object.

Now let us recall Newton's *law of inertia* : An object left to itself either remains at the same place or performs uniform motion (that is, motion along a straight line with constant speed).

Clearly, next to remaining stationary, uniform motion is the simplest type of motion. But we observe motion which is far more complicated than the uniform motion !

The reason behind the non-uniform motion is that objects are seldom left to themselves. Each object has a natural tendency to change the state of motion of any other object in its vicinity. Thus an object B tries, not only to change the position of a nereby object A but affects its velocity also. Thus, when the object A is moving in a region Ω in V, it experiences at each place in Ω a pull of certain strength in a certain direction. This pull is the force imparted on A by objects in the surroundings. In other words, matter in the surrounding generates a *field of forces*-more compactly expressed, a *force field*-which acts on the object A while moving in the region. It is this physical quantity— the force field-which deviates objects from their state of rest or uniform motion.

Do we not observe that there is a force field of one kind or other in every region around us ? Consequently we see no object lingering for too long a time in the same place (unless special efforts are made to maintain it in the same place). Instead, it moves around with varying rapidity and direction of its motion.

Indeed, motion is an important aspect of the physical world.

MECHANICS, the science of this all-pervasive motion is naturally a vast subject. (DYNAMICS is another, perhaps more popular term for it.) Consequently it is divided into a number of branches: Classical Mechanics, Hydromechanics, Statistical Mechanics, Quantum Mechanics $\cdots$. Each of these branches is characterized by (i) the type of objects, (ii) the nature of the forces acting on them, (iii) the nature of motion, (iv) the view-point from which the motion is studied.

Among all these branches, CLASSICAL MECHANICS is the most basic, most elementary and also the most important one because of its immediate utility.

In classical mechanics, we study the motion of a **mechanical system** consisting of a finite family of non-deforming material object. Each of these objects has a definite shape, moderate size, moderate amount of mass and is assumed to move with speed which is very small in comparison with the speed of light. The motion of the objects is considered to be **deterministic** in nature. (We will explain this term after a little while.)

In this book we study some elementary aspects of classical mechanics which are essentially based on the familiar Newton's laws of motion.

To begin with, let us consider the motion of a single object which is not deformable and which is neither too small (like an atom) nor too large (of astronomical scale.) We will refer to it as a **rigid body** or briefly a **body**.

Thus, a rigid body is a material object having moderate size, a definite shape, definite volume, a definite distribution of material content (i.e. its mass) in it and is *rigid* in the sense that it does not get appreciably deformed under the stresses and strains of its motion. In particular, the distance between any two points of the object remains unchanged throughout the course of its motion.

We list now the following elementary facts about the motion of such a rigid body. We call them *elementary facts* because, everyone of us has been observing them in the daily life and therefore no one denies their existence.

(I) Motion of a body is a combination of the following two types:

   (a) **Translational motion:** A body is transported from one place to another along its own course.

   (b) **Rotational motion:** While being transported, the body undergoes a change in its orientation in reference to other stationary objects surrounding it.

(II) The two types of motion, that is, the translation and rotation, are independent of each other.

(III) The smaller the size of an object, the lesser are the effects of its rotational motion.

(IV) The initial state of motion (that is, the place from where the object was set in motion together with its starting velocity) determines the entire future course of motion of the object.

The last property is at the foundation of our subject of discourse. Sometimes, especially in the more advanced treatment of the subject, this property is called **Newton's Principle of Determinacy**.

Thus, the nature of motion of a rigid body is what we have termed earlier *deterministic* in the sense that in the known physical surroundings (in particular, in a given force field,) it is possible to predict the motion of the body in terms of its initial state of motion.

Property (IV) is highly mathematical in nature !

To appreciate the above remark, all we need to do is to consider the parallel situation in Mathematics arising in the theory of ordinary differential equations (in the following referred to as the ODE). Recall only one result: The solution of a second order ODE is determined uniquely by the value of the solution and that of its first derivative, both evaluated at an initial value of the independent variable.

This striking analogy might have motivated Isaac Newton to formulate his well-known second law of motion in the form of the second order ODE:

$$\text{Mass} \times \text{Acceleration} = \text{Force} \qquad (1).$$

1.2   Concepts of a Particle

We have already noted the following points in the Introduction:

(1) Classical mechanics deals with motion of macroscopic rigid objects.
(2) The motion is governed by the second order ODE (1).

However, because of its recognizable size, a general rigid body is too complicated an object to apply equation (1) to it. Therefore we would like to apply some simplifying assumptions and get riddance of some of the non-essential physical realities of the body. The main intention in doing so is to extricate the translational motion from other complexities so that equation (1) becomes applicable.

We have noted earlier one more point, namely, the smaller the size of the object, the smaller are the effects of its rotational motion. Now, very often, the size of the object is so negligible in comparison with the extent of its orbit that we can altogether neglect the rotational motion of the object.

Now it should be clear to the reader that under the assumption of translational motion being predominant, we may consider all the mass of the body to be concentrated at a point of it and thereby arrive at the conceptual **point mass** or a **particle**. Evidently, equation (1) is applicable to such a concept of a point mass.

To summarize, when there is no appreciable rotational motion of a body, we may choose a convenient point in it and consider all the mass of the body to be concentrated at the point. The resulting (conceptual) point mass is considered to have essentially the same dynamical properties as those of the body. The force on the body is considered to act at the reference point. Also, the position, velocity, acceleration of the body are considered to be those of the representative point of the body. We now have the following informal definition of a particle:

Definition 1 A **particle** is a mathematical idealization of a body whose position at an instant can be represented by a point in the physical space V.

In the remaining part of this chapter, we will study the geometry of motion - the **kinematics** of the particle. In the next three chapters we will study the motion of a single particle in more detail, taking into consideration the cause of the motion, that is, the force field acting on it.

In **Chapter 5** we will consider the motion of a finite family of particles.

After being familiar with particle motion, we will consider the motion of an actual rigid body. This is accomplished by considering a rigid body as a kind of limit of a large family of particles. This is the subject matter of **Chapter 6**.

1.3   Motion of a Particle

In the proceeding section, we introduced the intellectual construct of a particle and explained how equation (1) of motion is applicable to it.

Obviously, a single particle is the simplest object to study its motion. We discuss it in more details now.

Let P be a particle and let m be its mass.

We study motion of P during an interval of time I. I could be its life-time or a part of it.

Suppose, at the instant $t \in I$, the particle P occupies the point $c(t) \in V$. Now the motion of P is the succession $\{c(t) : t \in I\}$ of points of V visited by P in the time interval I, the order of the succession being induced by the natural order of the time interval I.

Mathematically, this succession of points is a time-parameterized curve in V:

$$c : I \longrightarrow V \qquad (2)$$

We call curve (2) the **trajectory** of P.

Sometimes, we are interested only in the set of points visited by P but *not* how the set was time-traversed. We call the resulting set the **path** or the **orbit** of P. Thus, the path of P is only the set of points lying on the trajectory of P. It is the set obtained from the trajectory by divesting it of the time parameterization induced by the motion of P.

Clearly, the trajectory is the complete description of motion of a particle while a path of it is a partial account of the motion. We will explain more about this distinction at a latter stage.

We now exploit the geometry of the space V. Choosing a convenient point **O** we consider a frame of reference $\mathcal{F}$ with its origin at the chosen point O. Let $\vec{i}, \vec{j}$ and $\vec{k}$ be the usual unit vectors along the axes of $\mathcal{F}$. Now every point $A \in V$ can be represented by the vector $\vec{OA}$. In particular a point $c(t)$ on the trajectory (2) of P determines (and in turn is determined by) the vector from O to $c(t)$. We denote it by $\vec{r}(t)$. Thus the trajectory (2) of P now acquires the vectorial representation:

$$t \longmapsto \vec{r}(t) \qquad (3)$$

Moreover, let the coordinates of $c(t)$ with respect to $\mathcal{F}$ be $(x(t), y(t), z(t))$ Consequently we have

$$\vec{r}(t) = x(t)\vec{i} + y(t)\vec{j} + z(t)\vec{k}. \qquad (4)$$

We say : (a) $\vec{r}(t)$ is the **instantaneous position vector** of P and   (b) $(x(t), y(t), z(t))$ are its (instantaneous) **coordinates** with respect to the frame $\mathcal{F}$.

We want to study the motion of P using differential calculus. Obviously, the techniques of differential calculus are not applicable to the motion in its most general form. We must therefore restrict our study to a narrow class of motions to which differential calculus is applicable. To be more specific, we will assume that the motion of P is such that the function (3) representing the trajectory is at least twice continuously differentiable.

Note that the choice of the frame of reference $\mathcal{F}$ has enabled us to do the following two things:

(I) Identify the physical space V with the Euclidean space $\mathbb{R}^3$ by means of the function

$$V \longrightarrow \mathbb{R}^3; \qquad A \longrightarrow (x_A, y_A, z_A).$$

From now onwards, we will not make any distinction between the physical space V and the Euclidean space $\mathbb{R}^3$.

(II) Express the trajectory (2) in terms of the vector valued function (3) (with $\vec{r}(t) = x(t)\vec{i} + y(t)\vec{j} + z(t)\vec{k}$) or in terms of the real valued coordinate functions

$$t \longmapsto x(t), \quad t \longmapsto y(t), \quad t \longmapsto z(t) \tag{5}$$

Now, the requirement of being at least twice continuous differentiability of (4) is equivalent to the (usual) condition of being at least twice continuous differentiability of the real valued functions (5).

We now define the following terms:

Definition 2

(a) The derivative

$$\frac{d\vec{r}(t)}{dt} = \dot{\vec{r}}(t) = \dot{x}(i) + \dot{y}t\vec{j} + \dot{z}(t)\vec{k} \tag{6}$$

is the **instantaneous velocity** of P.

We denote it by $\vec{v}(t)$.

(b) $(\dot{x}(t), \dot{y}(t), \dot{z}(t))$ appearing in (6) are the **components**, of the velocity $\vec{v}(t)$ with respect to the frame $\mathcal{F}$.

(c) $\|\vec{v}(t)\| = \{\dot{x}(t)^2 + \dot{y}(t)^2 + \dot{z}(t^2\}^{\frac{1}{2}}$ is the **instananeous speed** of the particle. Another notation for it is $v(t)$.

Definition 3 The second derivative

$$\frac{d^2\vec{r}(t)}{dt^2} = \frac{d\vec{v}}{dt} = \ddot{\vec{r}}(t) = \ddot{x}(t)\vec{i} + \ddot{y}(t)\vec{j} + \ddot{z}(t)\vec{k} \tag{7}$$

is the **instantaneous acceleration of** P. It is denoted by $\vec{a}(t) \cdot (\ddot{x}(t).\ddot{y}(t), \ddot{z}(t))$ are its components with respect to $\mathcal{F}$. Magnitude of the acceleration $\vec{a}(t)$ is $\{\ddot{x}(t)^2 + \ddot{y}(t)^2 + \ddot{z}(t)^2\}^{\frac{1}{2}}$, and is denoted by $a(t)$. At this stage, we explain the distinction between the path and the trajectory of a particle. Recall, the trajectory of a particle is a curve $c : I \longrightarrow V$ while the path is the set $\{c(t) : t \in I\} \subset V$ freed of the time parameterization. In particular, notice that two particles may tread the same path in opposite directions.

Here is a simple illustration of the situation.

Let $\vec{e}$ be any unit vector. We consider the line segment L joining 0 to the point $A (\simeq \vec{e})$. We consider the motion of two particles P and Q during the time interval $[0, 1]$. (We may consider Q moving during the same interval of time namely $[0, 1]$ but during the *next day* (say) so·that they do not collide against each other.)

The trajectory of P is

$$c : [0, 1] \longrightarrow L; t \longmapsto t\vec{e}$$

and they of Q is

$$\tilde{c} : [0, 1] \longrightarrow L; t \longmapsto (1 - t)^2 \vec{e}.$$

Clearly, P and Q traverse the same path, namely, the line segment L. But the way the points on L are visited by P and Q are different:

(a) P moves from 0 to A while Q moves in the opposite direction.

(b) P moves with constant velocity $\vec{e}$ while Q moves with variable velocity $-2(1 - t)\vec{e}$.

(c) P has zero acceleration while Q has the constant (non-zero) acceleration $2\vec{e}$.

Thus, it is the time parameterization which describes the actual motion.

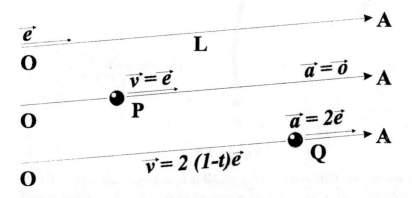

Figure 1.1

Now, we discuss some simple examples.

Example 1 A particle P moves with acceleration $= -g\vec{k}$, $g > 0$ being a constant. If its position at $t = 0$ is $A = (0, 0, h)$, $h > 0$, and its velocity $= \vec{i} + 2\vec{j}$, (i) find where and when the particle passes through the XOY plane and also (ii) describe the path traversed by the particle.

Solution: Let $c : [0, \infty) \longrightarrow \mathbb{R}^3; t \longmapsto \vec{r}(t) = x(t)\vec{i} + y(t)\vec{j} + z(t)\vec{k}$. be the trajectory of P. Then its instantaneous velocity and acceleration are $\vec{v}(t) = \dot{x}(t)\vec{i} + \dot{y}(t)\vec{j} + \dot{z}(t)\vec{k}$ and $\vec{a}(t) = \ddot{x}(t)\vec{i} + \ddot{y}(t)\vec{j} + \ddot{z}(t)\vec{k}$ respectively. Therefore, by the given data, we have

$$\ddot{x}(t)\vec{i} + \ddot{y}(t)\vec{j} + \ddot{z}(t)\vec{k} = -g\vec{k}.$$

Separating the components we get

$$\ddot{x}(t) \equiv 0, \quad \ddot{y}(t) \equiv 0 \ \ and \ \ \ddot{z}(t) \equiv -g.$$

Integrating these differential equations, we get

$$\dot{x}(t) = \alpha, \quad \dot{y}(t) = \beta \ \ and \dot{z}(t) = -gt + \gamma.$$

α, β and γ being some constants to be determined presently. Thus $\vec{v}(t) = \alpha\vec{i} + \beta j + (-gt + \gamma)\vec{k}$.

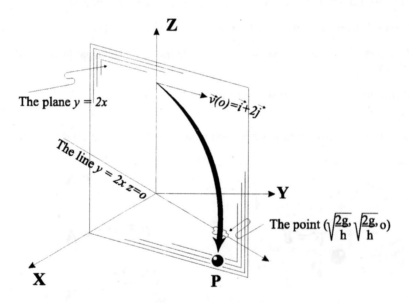

Figure 1.2

In particular, $\vec{v}(0) = \alpha\vec{i} + \beta\vec{j} + \gamma\vec{k}$. But it is given that $\vec{v}(0) = \vec{i} + 2\vec{j}$. Therefore $\alpha\vec{i} + \beta\vec{j} + \gamma\vec{k} = \vec{i} + 2\vec{j}$ which gives $\alpha = 1, \quad \beta = 2$ and $\gamma = 0$. Hence $\dot{x}(t) \equiv 1, \quad \dot{y}(t) \equiv 2$ and $\dot{z}(t) = -gt$. Integrating these equations, we get

$$x(t) \equiv t + a, \quad y(t) \equiv 2t + b, \quad z(t) = -\frac{gt^2}{2} + c$$

a, b, c being some constants. Therefore, $\vec{r}(t)$ is given by

$$\vec{r}(t) = (t + a)\vec{i} + (2t + b)\vec{j} + (-\frac{gt^2}{2} + c)\vec{k}.$$

In particular, $\vec{r}(0) = a\vec{i} + b\vec{j} + c\vec{k}$. But it is given that $\vec{r}(0) = h\vec{k}$. Therefore we must have $a\vec{i} + b\vec{j} + c\vec{k} = h\vec{k}$; which gives $a = 0$, $b = 0$ and $c = h$.

Hence the trajectory of the particle is

$$t \longmapsto t\vec{i} + 2t\vec{j} + (-\frac{gt^2}{2} + h)\vec{k}$$

for all $t \geq 0$.

Suppose, the particle passes through the XOY-plane at time $t = t_o$. Then we have

$$\vec{r}(t_o) = t_o\vec{i} + 2t_o\vec{j} + (-\frac{1}{2}gt_o^2 + h\vec{k})$$
$$= t_o\vec{i} + 2t_o\vec{j} + 0\vec{k},$$

since $\vec{r}(t_o)$ is a point in the XoY-plane. Comparing the coefficients of $\vec{k}$, we get

$$-\frac{1}{2}gt_o^2 + h = 0.$$

which gives $t_o = \sqrt{\left(\frac{2h}{g}\right)}$.

Therefore, point $\vec{r}(t_o)$ (of the XoY-plane through which the particle passes) is $\vec{r}\left(\sqrt{\left(\frac{2h}{g}\right)}\right) = \sqrt{\left(\frac{2h}{g}\right)}\left(\vec{i} + 2\vec{j}\right)$.

Note that $2x(t) = y(t)$, and $z(t) = -\frac{gt^2}{g} + h$ along the trajectory. Eliminating t from these equations, we get the trajectory

$$y = 2x, \quad z = h - \frac{gx^2}{2}.$$

Clearly it is the parabola $x^2 = \frac{2}{g}(h - z)$ in the plane $y = 2x$. □

Example 2 Particle P moves in a plane in such a way that its velocity $\vec{v}(t)$ and acceleration $\vec{a}(t)$ satisfy the equation

$$\vec{a}(t) = \alpha\vec{v}(t) + \vec{\beta} \cdots \qquad (*)$$

where α and β are constants. Prove that the direction of the acceleration vector does not change.

Solution Choose arbitrarily a point 0 on the path of the particle. Fix a frame of reference $\mathcal{F}$ by requiring

(a) the origin of $\mathcal{F}$ coincides with the point 0 chosen above.

(b) X-axis of $\mathcal{F}$ goes along $\vec{\beta}$.

(c) the plane of motion of P coincides with the XOY-plane of $\mathcal{F}$ (and consequently $z(t) \equiv 0$). Now $\vec{r}(t) = x(t)\vec{i} + y(t)\vec{j}$. Therefore we have

$$\vec{v}(t) = \dot{x}(t)\vec{i} + \dot{y}(t)\vec{j} \text{ and } \vec{a}(t) = \ddot{x}(t)\vec{i} + \ddot{y}(t)\vec{j}.$$

Let $\theta(t)$ be the angle between the X-axis and the acceleration vector $\vec{a}(t)$.

We must prove: $\theta(t)$ is independent of t.

Write $\vec{\beta} = \beta\vec{i}$ so that β is the magnitude of $\vec{\beta}$. Using (*) we get

$$\ddot{x}(t)\vec{i} + \ddot{y}(t)\vec{j} = \alpha \left[\dot{x}(t)\vec{i} + \dot{y}(t)\vec{j} \right] + \beta\vec{i}.$$

Equating the coefficients of $\vec{i}$ and $\vec{j}$ in the above equation, we get

$$\ddot{x}(t) = \alpha\dot{x}(t) + \beta \text{ and } \ddot{y}(t) = \alpha\dot{y}(t).$$

Dividing the first equation by second, we get

$$\frac{\ddot{x}(t)}{\ddot{y}(t)} = \frac{\alpha\dot{x}(t) + \beta}{\alpha\dot{y}(t)} \qquad (**)$$

or equivalently,

$$\frac{\ddot{x}(t)}{\alpha\dot{x}(t) + \beta} = \frac{\ddot{y}(t)}{\alpha\dot{y}(t)}.$$

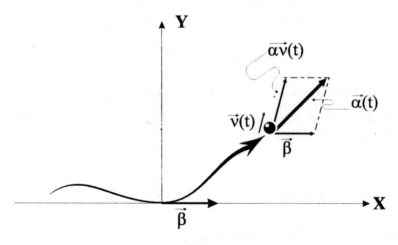

Figure 1.3

On integration, the above differential equation gives

$$\frac{1}{\alpha}\log(\alpha\dot{x}(t)+\beta)=\frac{1}{\alpha}\log\dot{y}(t)+\text{ constant}$$

and therefore

$$\frac{\alpha\dot{y}(t)}{\alpha\dot{x}(t)+\beta}=\text{ constant} \qquad (*\,*\,*)$$

Combining $(**)$ and $(***)$ we get $\frac{\ddot{y}(t)}{\ddot{x}(t)}$=constant. But $\frac{\ddot{y}(t)}{\ddot{x}(t)}=\tan\theta(t)$ and so the angle $\theta(t)$ remains constant. Being the angle between the acceleration vector $\vec{a}(t)$ and the X-axis, its constancy proves that the acceleration does not change its direction. □

Example 3 A wheel of radius a rolls with constant speed u along a straight track. Show that the speed of a point A on the rim of the wheel is given by

$$v(t)=u\left\{2-2\sin[\theta(t)+\alpha]\right\}^{\frac{1}{2}}$$

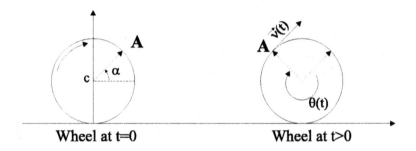

Wheel at t=0   Wheel at t>0

Figure 1.4

where α is the angle made by the radius vector $\vec{CA}$ with the straight track at $t=0$ and $\theta(t)$ is the angle turned by the radius vector $\vec{CA}$ about C in time t, C being the center of the wheel. Also, show that the acceleration of the point A has magnitude $\frac{u^2}{a}$ and is directed towards the center of the wheel.

Solution We take the coordinate frame $\mathcal{F}$ as shown in Fig. 1.4. (Since the motion is taking place in the vertical plane, we display only that plane as the XOY-plane of $\mathcal{F}$).

Clearly, the angular speed of the wheel is $\frac{u}{a}$.

Consequently, $\theta(t)=\frac{ut}{a}$ and the coordinates of A at time t are given by

$$x(t)=tu+a\cdot\cos(\frac{-tu}{a}+\alpha), y(t)=a\cdot\sin(\frac{-tu}{a}+\alpha), z(t)\equiv 0.$$

(As $z(t)\equiv 0$ we neglect it altogether in the following.)

Differentiating the coordinates with respect to t, we get

$$\dot{x}(t) = u + u \cdot \sin(\frac{-tu}{a} + \alpha), \dot{y}(t) = -u \cdot \cos(\frac{-tu}{a} + \alpha)$$

and $\dot{z}(t) \equiv 0$ (of course).

$$\ddot{x}(t) = -\frac{u^2}{a} \cdot \cos(\frac{-tu}{a} + \alpha), \ddot{y}(t) = -\frac{u^2}{a} \cdot \sin(\frac{-tu}{a} + \alpha).$$

Therefore, the speed of A at time t is

$$
\begin{aligned}
v(t) &= \left\{ \dot{x}(t)^2 + \dot{y}(t)^2 + \dot{z}(t)^2 \right\}^{\frac{1}{2}} \\
&= \left\{ u^2 - 2u^2 \sin(\frac{-tu}{a} + \alpha) + u^2 \sin^2(\frac{-tu}{a} + \alpha) \right. \\
&\quad \left. + u^2 \cos^2(\frac{-tu}{a} + \alpha) \right\}^{\frac{1}{2}} \\
&= u \left\{ 2 - \sin(\frac{-tu}{a} + \alpha) \right\}^{\frac{1}{2}} \\
&= u \left\{ 2 - \sin(\theta(t) + \alpha) \right\}^{\frac{1}{2}}
\end{aligned}
$$

Also, the acceleration of the point is:

$$
\begin{aligned}
\vec{a}(t) &= \ddot{x}(t)\vec{i} + \ddot{y}(t)\vec{j} \\
&= -\frac{u^2}{a} \left[\cos(\theta(t) + \alpha)\vec{i} + \sin(\theta(t) + \alpha)\vec{j} \right] \\
&= -\frac{u^2}{a} \vec{e}(t)
\end{aligned}
$$

where $\vec{e}(t)$ is the vector $\cos(\theta(t) + \alpha)\vec{i} + \sin(\theta(t) + \alpha)\vec{j}$ which is a unit vector along $\vec{CA}$. Thus the acceleration of A is $\frac{u^2}{a}$ directed along $\vec{AC}$. □

1.4 Motion in a Plane

In many situations, the particle remains confined to a single plane in V throughout the course of motion. We then say that the particle is performing *planar motion*.

For example, a planet in the solar system performs a planar motion. (We study planatory motion in **Chapter 4.**) In this section and the next, let us study some elementary aspects of the planar motion.

First, we choose a frame of reference $\mathcal{F}$ in such a way that the plane of motion of the particle coincides with the XOY-plane of $\mathcal{F}$. Now, the position vector $\vec{r}(t)$ of P is a vector in the XOY-plane (it has no Z-component)

$$\vec{r}(t) = x(t)\vec{i} + y(t)\vec{j} \tag{8}$$

Differentiating equation (8) twice, we get

$$\vec{v}(t) = \dot{x}(t)\vec{i} + \dot{y}(t)\vec{j} \tag{9}$$

$$\vec{a}(t) = \ddot{x}(t)\vec{i} + \ddot{y}(t)\vec{j} \tag{10}$$

Thus, the right choice of $\mathcal{F}$ enables us to eliminate the Z-component from many of the vectorial quantities such as position vector, velocity, acceleration etc.

Example 4 A particle moves along an elliptical path in a plane with constant speed. Find an expression for the magnitude of the acceleration as a function defined on the ellipse. At what points the acceleration has maximum and minimum magnitudes ?

Solution We choose the frame $\mathcal{F}$ in such a way that (in addition to its XOY-plane coinciding with the plane of the ellipse,) the origin is at the center of the ellipse and the axes of $\mathcal{F}$ are along the axes of the ellipse. Let the equation of the ellipse be $\frac{x^2}{a^2} + \frac{y^2}{b^2} = 1$. As the particle is moving along the ellipse, its coordinates $(x(t), y(t))$ satisfy the equation

$$\frac{x(t)^2}{a^2} + \frac{y(t)^2}{b^2} = 1 \tag{*}$$

In the remaining part of the solution, we will write (x, y) instead of $(x(t), y(t))$ to simplify the notations. Differentiating (*) with respect to t, we get

$$\frac{x\dot{x}}{a^2} + \frac{y\dot{y}}{b^2} = 0$$

and therefore,

$$\dot{y} = -\frac{b^2}{a^2} \cdot \frac{x}{y} \cdot \dot{x} \tag{**}$$

at all points of the ellipse except where $y = 0$.

Let v be the speed of the particle. Then we have

$$\dot{x}^2 + \dot{y}^2 = v^2 \tag{***}$$

Substituting (**) in (***), we get

$$\dot{x}^2 \left(1 + \frac{b^4}{a^4} \cdot \frac{x^2}{y^2} \right) = v^2 \tag{4*}.$$

Differentiating this equation, we get

$$2\dot{x} \cdot \ddot{x} \left(1 + \frac{b^4}{a^4} \frac{x^2}{y^2} \right) + 2\dot{x}^2 \frac{b^4}{a^4} \cdot \frac{1}{y^4} \left[y^2 \cdot x \cdot \dot{x} - x^2 \cdot y \cdot \dot{y} \right] = 0.$$

Simplifying and rewriting we get

$$\ddot{x} \left[1 + \frac{b^4}{a^4} \frac{x^2}{y^2} \right] + \dot{x} \frac{b^4}{a^4} \cdot \frac{x \cdot y}{y^4} \left[y\dot{x} - x\dot{y} \right] = 0.$$

Using (∗∗) to simplify the above, we get

$$\ddot{x}\left[1 + \frac{b^4}{a^4}\frac{x^2}{y^2}\right] + \dot{x}^2\frac{b^6}{a^4}\frac{x}{y^4} = 0.$$

and therefore,

$$\ddot{x} \doteq \frac{-b^6 x\dot{x}^2}{a^4 y^4\left(1 + \frac{b^4}{a^4}\frac{x^2}{y^2}\right)} = \frac{-b^6 v^2 x}{a^4 y^4\left(1 + \frac{b^4}{a^4}\frac{x^2}{y^2}\right)^2}, \text{ using } (4\ast)$$

$$\doteq \frac{-a^4 b^6 v^2 x}{(a^4 y^2 + b^4 x^2)^2}$$

Thus,

$$\ddot{x} = \frac{-a^4 b^6 v^2 x}{(a^4 y^2 + b^4 x^2)^2} \qquad (5\ast)$$

Next, differentiating (∗ ∗ ∗) we get $\dot{x}\cdot\ddot{x} + \dot{y}\cdot\ddot{y} = 0$ and, consequently

$$\ddot{y} = -\frac{\dot{x}}{\dot{y}}\ddot{x} \qquad (6\ast)$$

Let $\alpha(x,y)$ denote the magnitude of the acceleration of the particle at a point (x,y) on the ellipse. Then we have

$$\begin{aligned}
\alpha^2 &= \ddot{x}^2 + \ddot{y}^2 \\
&= \left(1 + \frac{\dot{x}^2}{\dot{y}^2}\right)\ddot{x}^2, \text{ using } (6\ast) \\
&= \left(\frac{b^4 x^2 + a^4 y^2}{b^4 x^2}\right)\ddot{x}^2 \text{ using } (\ast\ast) \\
&= \frac{(a^4 y^2 + b^4 x^2)}{b^4 x^2}\frac{a^8 b^{12} v^4 x^2}{a^4 y^2 + b^4 x^2)}^4, \text{ using } (5\ast) \\
&\doteq \frac{a^8 b^8 v^4}{(a^4 y^2 + b^4 x^2)^3}
\end{aligned}$$

Hence $\alpha = \alpha(x,y) = \frac{a^4 b^4 v^2}{(a^4 y^2 + b^4 x^2)^{3/2}}$.

Clearly, $\alpha(x,y)$ will be maximum (minimum) where the polynomial $(a^4 y^2 + b^4 x^2)$ attains its minimum (maximum) value on the ellipse.

Using the parameterization $x = a\cdot\cos\theta, y = b\sin\theta$, we can find the maximum and the minimum values of $\alpha(x,y)$ on the ellipse. We leave this part of the solution for the reader.   □

1.5 Polar Coordinates

We continue our study of the motion in a plane. Let us denoted the plane of motion by $\sum$.

Often, there is a point of $\sum$ around which the particle gyrates without passing through it. For example, this is how it is in the case of motion of planet in the solar system.

In such a situation, we choose the frame $\mathcal{F}$ in such a way that (in addition to the property that the plane of motion coinciding with the XOY-plane of $\mathcal{F}$), the origin of $\mathcal{F}$ coincides with the distinguished point.

We can of course use the (x, y) coordinates provided by the frame $\mathcal{F}$. However, there is another pair of (non-Cartesian) coordinates which turn out to be better suited to the gyrational motion of the particle about the origin of $\mathcal{F}$. These new coordinates are called the **polar coordinates** about the origin. We describe them below:

First (in this context), we consider the positive X-axis of $\mathcal{F}$ and call it the **initial direction** for the polar coordinates.

Now, let A be any point of $\sum \backslash \{0\}$.

Suppose

(i) $\angle AOX$ has measure θ radians and

(ii) dist $(0, A) = r$.

Then the pair (r, θ) is said to be the **polar coordinates** of the point A.

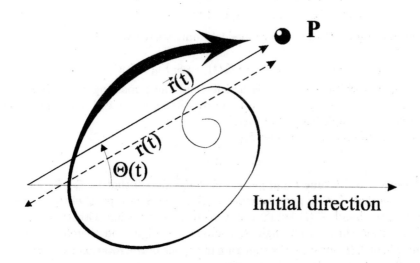

Figure 1.5

Thus, we have defined polar coordinates of *any* point in the punctured plane $\sum \backslash \{0\}$. Note that the Cartesian coordinates (x, y) and the polar coordinates (r, θ) determine each other uniquely:

$$x = r \cdot \cos\theta, y = r \cdot \sin\theta; r = \sqrt{x^2 + y^2} \text{ and } \theta = \tan^{-1}\left(\frac{y}{x}\right).$$

Comming back to the discussion of the motion of a particle P in $\sum \backslash \{0\}$, let $(r(t), \theta(t))$ be the polar coordinates of the instantaneous position $c(t)$ of P. The trajectory $c : I \longrightarrow \sum \backslash \{0\}$ of P thus gives rise to the two differentiable functions:

$$I \longrightarrow [0, \infty); \quad t \longmapsto r(t)$$

and

$$I \longrightarrow [0, 2\pi); \quad t \longmapsto \theta(t).$$

the first describing the distance of P from the center of gyration (i.e. the point 0) and the second describing the angle subtended by the position vector $\vec{r}(t)$ of P with the initial direction.

As we plan to switch over from the Cartesian coordinates to the Polar coordinates, we need express the velocity and the acceleration of the particle in terms of the polar coordinates and their time derivatives. Towards this goal, we associate with the motion of P two mutually perpendicular unit vectors $\vec{e}_r(t)$ and $\vec{e}_\theta(t)$ as follows:

(i) The unit vector $\vec{e}_r(t)$ in $\sum$ is parallel with $\vec{r}(t)$ and points from 0 towards P (see Fig. 1.6(b)) on the next page. Thus $\vec{r}(t) = r(t)\vec{e}_r(t)$.

(ii) $\vec{e}_\theta(t)$ is the unit vector in $\sum$ perpendicular to $\vec{e}_r(t)$ in such a way that the ordered triple $\{\vec{e}_r(t), \vec{e}_\theta(t), \vec{k}\}$ is *right handed* or equivalently stated, $\vec{e}_r(t) \times \vec{e}_\theta(t) \equiv \vec{k}$. (see Fig. 1.6(c)).

The motion of P thus determines two unit vector valued functions in $\sum$:

$$t \longmapsto \vec{e}_r(t) \text{ and } t \longmapsto \vec{e}_\theta(t) \quad \text{see Fig 1.6(a)}.$$

In the remaining part of this article (and the remaining part of the book also,) we will simplify our notation by witting $\vec{e}_r$ instead of $\vec{e}_r(t)$ and similarly, we write $\vec{e}_\theta$ instead of $\vec{e}_\theta(t)$.

Now we differentiate the identity

$$\vec{e}_r \cdot \vec{e}_r \equiv 1$$

with respect to the time t and get $\dot{\vec{e}}_r \cdot \vec{e}_r + \vec{e}_r \cdot \dot{\vec{e}}_r = 0$ that is, $\dot{\vec{e}}_r \cdot \vec{e}_r = 0$. This shows that $\dot{\vec{e}}_r(t)$ is perpendicular to $\vec{e}_r(t)$ and hence parallel to $\vec{e}_\theta(t)$. On the other hand, $\vec{e}_r(t)$ being a unit vector, its variation with time can be its rotation through an angle $\theta(t)$ about the origin and therefore, the magnitude of $\dot{\vec{e}}_r(t)$ must be the time rate of change of the rotation $t \longmapsto \theta(t)$ that is $\dot{\theta}(t)$. Thus, we have the identity

$$\dot{\vec{e}}_r = \dot{\theta}\vec{e}_\theta \qquad (11)$$

Similarly, it can be shown that

$$\dot{\vec{e}}_\theta = -\dot{\theta}\vec{e}_r \qquad (12)$$

(Note that the negative sign on the right hand side of (12) appears because $\vec{e}_\theta$ is ahead of $\vec{e}_r$ in the counterclockwise sense of rotations.)

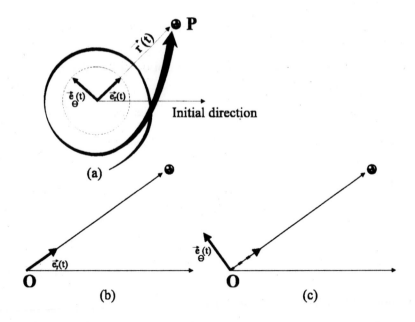

Figure 1.6

Now differentiating the identity $\vec{r}(t) = r(t)\vec{e}_r(t)$, we get

$$
\begin{aligned}
\dot{\vec{r}} &= \dot{r}\vec{e}_r + r\dot{\vec{e}}_r \\
&= \dot{r}\vec{e}_r + r\dot{\theta}\vec{e}_\theta
\end{aligned}
$$

Thus, we have the identity

$$\dot{\vec{r}}(t) = \dot{r}(t)\vec{e}_r(t) + r(t)\dot{\theta}(t)\vec{e}_\theta(t) \tag{13}$$

$\dot{r}(t)$ is called the **radial component** and $r(t)\dot{\theta}(t)$ the **transversal component** of $\dot{\vec{r}}(t)$. The pair $(\dot{r}(t), r(t)\dot{\theta}(t))$ is the **polar decomposition** of the velocity of P.

We now differentiate (13) to get the acceleration

$$
\begin{aligned}
\vec{a}(t) &= \frac{d}{dt}[\dot{r}\vec{e}_r + r\dot{\theta}\vec{e}_\theta] \\
&= \ddot{r}\vec{e}_r + \dot{r}\dot{\vec{e}}_r + \dot{r}\dot{\theta}\vec{e}_\theta + r\ddot{\theta}\vec{e}_\theta + r\dot{\theta}\dot{\vec{e}}_\theta \\
&= \ddot{r}\vec{e}_r + \dot{r}\dot{\theta}\vec{e}_\theta + \dot{r}\dot{\theta}\vec{e}_\theta + r\ddot{\theta}\vec{e}_\theta - r\dot{\theta}^2\vec{e}_r, \text{ using } (11), (12). \\
&= (\ddot{r} - r\dot{\theta}^2)\vec{e}_r + (2\dot{r}\dot{\theta} + r\ddot{\theta})\vec{e}_\theta.
\end{aligned}
$$

Thus we have

$$\vec{a}(t) = \left[\ddot{r}(t) - r(t)\dot{\theta}^2(t)\right]\vec{e}_r(t) + \left[2\dot{r}(t)\dot{\theta}(t) + r(t)\ddot{\theta}(t)\right]\vec{e}_\theta(t) \tag{14}.$$

Again $\ddot{r}(t) - r(t)\dot{\theta}^2$ is the **radical component** and $2\dot{r}\dot{\theta}(t) + r(t)\ddot{\theta}(t)$ is the **transversal component** of the acceleration.

Example 5 A particle moves in a plane, the position of it being given in terms of its polar coordinates by

$$r(t) = bt^2, \theta(t) = ct, \ b \text{ and } c \text{ being constants.}$$

Find polar decompositions of its velocity and acceleration as functions of the time t.

Solution We have

$$
\begin{aligned}
\dot{r}(t) &= 2bt \\
\ddot{r}(t) &= 2b \\
\dot{\theta} &= c \\
\ddot{\theta}(t) &= 0
\end{aligned}
$$

Therefore, the required polar decompositions are given by

$$
\begin{aligned}
\text{(I) \quad velocity} &= \left(\dot{r}(t), r(t)\dot{\theta}(t) \right) \\
&= \left(2bt, \quad bct^2 \right) ; \\
\text{(II) \quad acceleration} &= \left(\ddot{r}(t) - r(t)\dot{\theta}(t)^2, 2\dot{r}(t)\dot{\theta}(t) + r(t)\ddot{\theta}(t) \right) \\
&= \left(2b - bt^2 \cdot c^2 \quad 4bt \cdot c + bt^2 \cdot 0 \right) \\
&= \left(2b - bc^2 t^2, \quad 4bct \right)
\end{aligned}
$$

□

Example 6 An aircraft pursues a straight course with constant speed u and is being chased by a guided missile moving with constant speed $2u$. The missile is fitted with a homing device to ensure that its motion is always directed towards the target. Initially, the missile is at right angles to the course of the aircraft and at a distance R from it.

Find the pursuit curve of the missile *relative* to the aircraft, taking the course of the aircraft as the initial direction. Also find the time taken by the missile to reach the aircraft.

Solution We take the (moving) frame $\mathcal{F}$ as shown in the figure. (The z-axis is not shown, it goes into the plane of the paper.)

At an instant $t > 0$, the polar decomposition of the position vector $\vec{r}(t)$ is:

$$(-(2u + u\cos\theta), u\sin\theta)$$

Thus, we have

$$(\dot{r}, r\dot{\theta}) = (-(2u + u\cos\theta), u\sin\theta)$$

which gives

$$(i) \quad \dot{r} = -u(2 + \cos\theta)$$

$$(ii) \quad r\dot{\theta} = u\sin\theta.$$

Dividing (i) by (ii) and eliminating t, we get

$$\frac{1}{r}\frac{dr}{d\theta} = \frac{-(2 + \cos\theta)}{\sin\theta}.$$

On separating the variables the above becomes

$$\frac{dr}{r} = \frac{-(2 + \cos\theta)}{\sin\theta}d\theta.$$

Integrating it, we get

$$\frac{A}{r} = \tan^2\frac{\theta}{2}\sin\theta, A \text{ being a constant.}$$

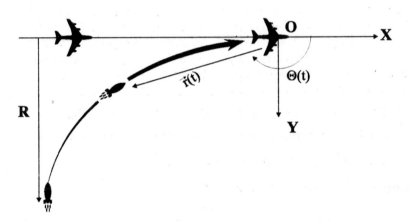

Figure 1.7

Now when $t = 0, \theta = \frac{\pi}{2}$ and $r = R$. Substitution of these values in the above equation gives $A = R$. Therefore, the required equation is $\frac{R}{r} = \tan^2\frac{\theta}{2}\sin\theta$. Now, when the missile reaches the aircraft, we have $r = 0$ and $\theta = \pi$. Also,

$$\dot{\theta} = \frac{u\sin\theta}{r}$$

$$= \frac{u\sin^2\theta \cdot \tan^2(\frac{\theta}{2})}{R}$$

$$= \frac{4u}{R}\sin^4(\frac{\theta}{2})$$

Therefore,

$$dt = \frac{R}{4u}\csc^4(\frac{\theta}{2})d\theta.$$

that is,

$$dt = \frac{R}{4u}(1 + \cot^2(\frac{\theta}{2}))\csc^2(\frac{\theta}{2})d\theta. \qquad (*)$$

Let τ denote the time taken by the missile to reach the aircraft. Integrating $(*)$ from $t = 0$ to $t = \tau$ (which corresponds to $\theta = \frac{\pi}{2}$ to $\theta = \pi$), we get

$$\tau = \int_0^{\tau} dt = \frac{4}{u}\int_{\pi/2}^{\pi}(1 + \cot^2(\frac{\theta}{2}))\cos ec^2(\frac{\theta}{2})d\theta = \frac{2}{3}\frac{R}{u}.$$

Thus the required time is $\frac{2R}{3u}$.   □

Of course our discussion above regarding the use of polar coordinates *does not* suggest that we should use *only* the polar coordinates to investigate the planar motion. The usual Cartesian coordinates (x, y) are also useful (especially, when we are not interested in the rotational aspect of the motion). The following example (in which the situation is similar to that of **Example 6**) illustrates this point.

Example 7 A flying target moves with uniform speed v along the line $y = h$ in the XOY-plane (where the X-axis is horizontal and the Y-axis .vertical.) At time $t = 0$, the target is located at the point $(0, h)$ and a guided missile P starts from the origin and moves with constant speed $2v$ in such a way that its velocity is always directed towards the target. Show that at time $t > 0$, if the coordinates of P are $(x(t), y(t))$ $[= (x, y)]$ and if $\frac{dx}{dy} \equiv q$; then $q(h - y) = vt - x$ and $1 + q^2 = 4v^2\left(\frac{dt}{dy}\right)^2$. By differentiating the first equation, with respect to y and eliminating $\frac{dt}{dy}$, derive a differential equation relating q and y. Deduce $\left[q + (1 + q^2)^{\frac{1}{2}}\right]^2 = \frac{h}{(h-y)}$.

Solution Let the target be denoted by T.

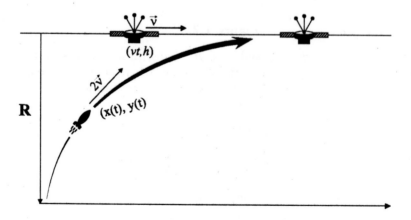

Figure 1.8

Clearly, the position of T at time t is (vt, h).

Also, the missile P is approaching T along the trajectory $t \longmapsto (x(t), y(t))$. If the tangent to the trajectory-which is also the instantaneous direction of motion of P-makes an angle $\theta(t)$ with the X-axis, then we have

$$\frac{1}{q} = \frac{dy}{dx} = \frac{\dot{y}(t)}{\dot{x}(t)} = \tan \theta(t).$$

Consequently, we have

$$\frac{h - y(t)}{[(h - y(t))^2 + (vt - x(t))^2]^{\frac{1}{2}}} = \sin \theta(t).$$

Rewriting it in the form

$$(h - y)^2 \csc^2 \theta = (h - y)^2 + (vt - x)^2$$

or equivalently

$$(h - y)^2 \left[1 + \cot^2 \theta \right] = (h - y)^2 + (vt - x)^2$$

which gives

$$(h - y)^2 \cot^2 \theta = (vt - x)^2$$

and hence

$$q(h - y) = (vt - x) \qquad\qquad (*)$$

Also, the instantaneous velocity of P has components $(\dot{x}(t), \dot{y}(t))$ while its magnitude is $2v$. Therefore, we must have

$$
\begin{aligned}
4v^2 &= \dot{x}(t)^2 + \dot{y}(t)^2 \\
&= \dot{y}(t)^2 \left[1 + \left(\frac{\dot{x}(t)}{\dot{y}(t)} \right)^2 \right] \\
&= \dot{y}(t)^2 \left[1 + q^2 \right]
\end{aligned}
$$

This gives

$$1 + q^2 = 4v^2 \left(\frac{dt}{dy} \right)^2 \qquad\qquad (**)$$

We differentiate $(*)$ with respect to y to get

$$\frac{dq}{dy}(h - y) - q = v \frac{dt}{dy} \div q$$

and hence

$$\frac{dq}{dy}(h - y) = v \frac{dt}{dy} = \frac{v(1 + q^2)^{\frac{1}{2}}}{2v} \quad \text{using } (**)$$

which gives the differential equation $\frac{dq}{dy} = \frac{(1+q^2)^{\frac{1}{2}}}{2(h-y)}$ which is the required differential equation between q and y. We rewrite it in the differential form

$$\frac{dq}{(1+q^2)^{\frac{1}{2}}} = \frac{dy}{2(h-y)}.$$

Integrating it, we get

$$\log\left\{q + (1+q^2)^{\frac{1}{2}}\right\} = \log\sqrt{(h-y)} + C \qquad (***)$$

C being the constant of integration.

Now when $y = 0, q = 0$ which gives $C = \log\sqrt{h}$. Substituting this value of C in $(***)$, we get

$$\left(q + (1+q^2)^{\frac{1}{2}}\right)^2 = \frac{h}{h-y}.$$

□

Example 8 A particle moves in a plane in such a way that its velocity components along and perpendicular to its position vector are λr and $\mu\theta$ respectively, λ and μ being some constants. Find the polar equation of the path of the particle and show that the polar decomposition of the acceleration of P is given by

$$\left(\lambda^2 r - \frac{\mu^2 \cdot \theta^2}{r}, \mu\theta\left[\lambda + \frac{\mu}{r}\right]\right).$$

Solution The polar decomposition of the velocity of P is $(\lambda r, \mu\theta)$ and, therefore we have

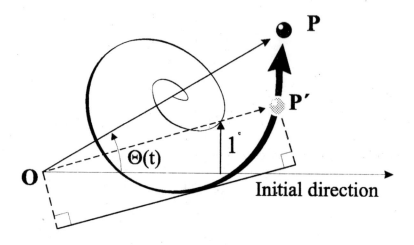

Figure 1.9

$$(\dot{r}, r\dot{\theta}) = (\lambda r, \mu\theta)$$

which gives: $\dot{r} = \lambda r$ and $r\dot{\theta} = \mu\theta$ or equivalently, $\dot{\theta} = \frac{\mu\theta}{r}$. Differentiating these equations with respect to time, we get

$$
\begin{aligned}
\ddot{r}(t) &= \frac{d}{dt}(\dot{r}(t)) = \frac{d}{dt}(\lambda r(t)) \\
&= \lambda \dot{r}(t) \\
&= \lambda^2 \cdot r(t)
\end{aligned}
$$

Also

$$
\begin{aligned}
\ddot{\theta}(t) &= \frac{d}{dt}(\dot{\theta}(t)) = \frac{d}{dt}\left(\mu\frac{\theta(t)}{r(t)}\right) \\
&= \mu\left[\frac{r(t)\dot{\theta}(t) - \theta(t)\dot{r}(t)}{r(t)^2}\right] \\
&= \mu\left[\frac{r(t)\mu\frac{\theta(t)}{r(t)} - \lambda r(t)\theta(t)}{r(t)^2}\right] \\
&= \frac{\mu^2\theta(t)}{r(t)^2} - \frac{\lambda\mu\theta(t)}{r(t)}
\end{aligned}
$$

Now, the radical component of the acceleration is

$$\ddot{r} - r\dot{\theta}^2 = \lambda^2 r - \frac{r\mu^2\theta^2}{r^2} = \lambda^2 r - \frac{\mu^2\theta^2}{r}. \tag{$*$}$$

while the transversal component of the acceleration is

$$
\begin{aligned}
2\dot{r}\dot{\theta} + r\ddot{\theta} &= \frac{2\lambda r\mu\theta}{r} + r\left[\frac{\mu^2\theta}{r^2} - \frac{\lambda\mu\theta}{r}\right] \\
&= 2\lambda\mu\theta + \frac{\mu^2\theta}{r} - \lambda\mu\theta \\
&= \mu\theta\left[\frac{\mu}{r} + \lambda\right]
\end{aligned}
\tag{$**$}
$$

Putting together $(*)$ and $(**)$ we get the required polar decomposition of acceleration

$$\left(\lambda^2 r - \frac{\mu^2\theta^2}{r}, \mu\theta\left[\lambda + \frac{\mu}{r}\right]\right).$$

Also we have $\frac{dr}{d\theta} = \frac{\dot{r}}{\dot{\theta}(t)} = \frac{\lambda r^2}{\mu\theta}$ which gives $\frac{dr}{r^2} = \frac{\lambda}{\mu}\frac{d\theta}{\theta}$. Integrating this, we get

$$\frac{-1}{r} + C = \frac{\lambda}{\mu}\log\theta, C \text{ being a constant.}$$

Let now b be the distance of the particle from 0 when $\theta = 1$ radian. Then we get $C = \frac{1}{b}$; and so, the equation of the orbit is

$$\frac{1}{b} = \frac{1}{r} + \frac{\lambda}{\mu} \, \log \theta.$$

$\square$

Example 9 A particle P moves in a plane in such a way that (i) the component of its velocity along a fixed direction is u and (ii) the transversal component of the velocity is $v - u \sin \theta$. Prove that the orbit of the particle is a conic section having eccentricity $\frac{u}{v}$.

Solution We take X-axis along the given fixed direction. Now, the polar decomposition of the velocity $\dot{\vec{r}}(t)$ is given by

$$\dot{\vec{r}}(t) = (u \cos \theta, v - u \sin \theta) = (\dot{r}, r\dot{\theta}).$$

Thus

$$\dot{r} = u \cos \theta \qquad\qquad (*)$$

and

$$r\dot{\theta} = v - u \sin \theta \qquad\qquad (**)$$

Dividing ($*$) by ($**$) we get

$$\frac{1}{r}\frac{dr}{d\theta} = \frac{u \cos \theta}{v - u \sin \theta}.$$

Separating the variables and integrating, we get

$$\log(r) = -\log(v - u \sin \theta) + \text{ constant}.$$

or equivalently, $\log\{(r \cdot (v - u \sin \theta) = \text{ constant}$. Therefore, $r(v - u \sin \theta) = \alpha$ where α is a constant. We replace θ by $\theta + \pi/2$, that is, we rotate the frame $\mathcal{F}$ about the origin through an angle $\frac{3\pi}{2}$. Writing e for $\frac{u}{v}$, the above equation becomes:

$$1 + e \cdot \cos \theta = \frac{\alpha}{r}$$

which is a standard equation of a conic section.

$\square$

1.6   Equation of Motion

In this article, we study equation (1) in more details.

We suppose, we have chosen a *stationary* frame of reference $\mathcal{F}$ once and for ever. All the vector quantities pertaining to the motion of a particle P will be understood to be taken in reference to $\mathcal{F}$.

As P is moving with velocity $\vec{v}(t)$, all the material content m of it that is its total mass is moving with velocity $\vec{v}(t)$. It is therefore natural to consider the product $m \cdot \vec{v}(t)$. In the heuristic sense, it is the *amount of motion* packed within P.

Recall now Newton's second law of motion. "The time rate of change of *motion* is proportional to he motive force impressed and $\cdots$." The single world *motion* appearing here is what we described above the *amount of motion*. We introduce (and use it throughout) yet another term for it in the following definition:

Definition 4 The vector quantity $m \cdot \vec{v}(t)$ is the instantaneous **linear momentum** of the particle.

We denote it by $\vec{p}(t)$.

Let us consider the concept of the *motive force* now. Often, the physical situation is such that instead of all of V, the particle can move only in a certain region $\Omega \subset V$. Thus a particle may be *constrained* to slide along a curve c in V, it may be made to roll on a surface $\sum$ in V, or it may be confined to the interior W of a container. Of course, Ω can also be the whole of V.

Consider the motion of P in Ω.

Let A be a point in Ω. While at A, the particle experiences a pull of certain strength in a certain direction. (This pull is the interaction of P with the surrounding matter; this interaction was explained in the **Introduction**.) This pull is thus a vector, we denote it by $\vec{F}(A)$. We say that $\vec{F}(A)$ is the **force** acting on P at A. The presence of a force at every point of Ω gives rise to a vector valued function

$$F : \Omega \longrightarrow \mathbb{R}^3 ; A \longmapsto \vec{F}(A)$$

Definition 5 The map $F : \Omega \longrightarrow \mathbb{R}^3$ is called the **force field** acting on P.

Now that we are familiar with the concepts of (a) the linear momentum of P and (b) the force field acting on it, we cast Newton's second law of motion in the form:

$$\frac{d}{dt}\vec{p}(t) = \vec{F}(\vec{r}(t)) \tag{15}$$

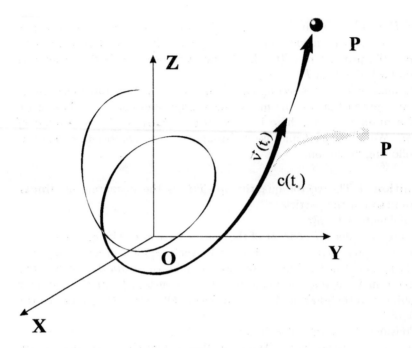

Figure 1.10

Moreover, since the mass of P is not affected by the motion, we get

$$\frac{d}{dt}\vec{p}(t) = m\frac{d\vec{v}}{dt}(t) = m\ddot{\vec{r}}(t) = m\vec{a}(t)$$

and so, equation (16) takes the more familiar form

$$m\ddot{\vec{r}}(t) = \vec{F}(\vec{r}(t)) \tag{16}$$

described in equation (1).

In most of what is to follow, we will use (16) as the equation of motion of the particle.

At this stage, let us consider the following aspect of the motion of P.

Suppose P is traversing the curve $c : I \longrightarrow \Omega; \quad t \longmapsto \vec{r}(t)$ under the influence of the force field F in Ω. Suppose a physical change takes places in the surroundings because of which the force field ceases to act on P from the moment $t = t_o$ onwards. Therefore, the equation of motion of P becomes

$$m\ddot{\vec{r}}(t) = 0$$

for all $t \geq t_o$ which implies $\frac{d}{dt}(\dot{\vec{r}}(t)) = 0$ (since $m > 0$) that is the curve $t \longmapsto \dot{\vec{r}}(t)$ is a constant for all $t \geq t_o$. Therefore,

$$\frac{d\vec{r}}{dt}(t) \equiv \dot{\vec{r}}(t_o). \tag{17}$$

Integrating (17) we get the trajectory

$$\vec{r}(t) = \vec{r}(t_o) + (t - t_o)\dot{\vec{r}}(t_o) \quad \forall t \geq t_o. \tag{18}$$

Thus we have proved the following

Proposition 1 Suppose, the particle is moving along a trajectory $c : I \rightarrow \Omega$ under the influence of a force field F. If the force field F stops acting on P from $t = t_o$ onwards and if there is no other force field acting on P, then the particle leaves the trajectory c at the point $c(t_o)$ and moves along the straight line tangential to c at $c(t_o)$ with (constant) velocity $\dot{\vec{r}}(t_o)$. Consequently, the trajectory of P (after leaving c at $c(t_o)$) is given by the curve (18).

The force fields considered so far were of the special type. The force on P (while at a point A) would depend only on the point A.

There are more general situations in which the force acting on P at A depends not only on the position of P, but on its velocity and the time as well.

$$F = \vec{F}(\vec{r}, \vec{v}, t).$$

Then the equation of motion of the particle is

$$m\ddot{\vec{r}}(t) = \vec{F}(\vec{r}(t), \dot{\vec{r}}, t). \tag{19}$$

As this book is intended to be only an introduction to the subject, we have only mentioned the most general form (19) of the equation of motion. We will solve it only in the following two types of force fields

Type I: $\vec{F}(\vec{r}, \vec{v}, t) = \vec{F}(t).$

Thus the force acting on P is independent of the velocity of P, and at a time t it is the same at all the places. Of course it may change with time. In other words we are dealing with a velocity independent, uniform and time dependent force field.

We integrate the equation of motion

$$m\ddot{\vec{r}}(t) = \vec{F}(t).$$

Suppose at $t = t_o$, the position of P is $\vec{r_o}$ and the velocity is $\vec{u}$. We integrate the equation

$$m\ddot{\vec{r}}(t) = \vec{F}(t)$$

to get

$$m\dot{\vec{r}}(t) = m\vec{u} + \int_{t_o}^{t} \vec{F}(s)ds.$$

and then $m\vec{r}(t) = m\vec{r_o} + m(t - t_o)\vec{u} + \int_{t_o}^{t} \left\{ \int_{t_o}^{s} \vec{F}(r)dr \right\} ds.$ Therefore the trajectory of the particle is given by

$$t \longmapsto r(t) = \vec{r_o} + (t - t_o)\vec{u} + \frac{1}{m}\int_{t_o}^{t} \left\{ \int_{t_o}^{s} \vec{F}(r)dr \right\} ds.$$

Type II : In all the remaining chapters of this book we will consider force fields of the type

$$F = \vec{F}(\vec{r}, \vec{v}t) = \vec{F}(\vec{r}).$$

We conclude this chapter by considering some more examples.

Example 10 A particle P having mass m is acted upon by the uniform, time dependent force field

$$\vec{F}(t) = 12\vec{a} + (cost)\ \vec{b}$$

$\vec{a}$ and $\vec{b}$ being two constant vectors.

Suppose, the particle was set in motion at time $t = t_o$ from a point A having position vector $\vec{\alpha}$ with the velocity $\vec{\beta}$. Determine the trajectory of the particle.

Solution The equation of motion is

$$m\ddot{\vec{r}}(t) = 12\vec{a} + (\cos t) \cdot \vec{b}.$$

Integrating it, we get

$$m\dot{\vec{r}}(t) = 12t\vec{a} + (\sin t)\vec{b} + \vec{u}$$

$\vec{u}$ being a constant vector to be determined.

Now it is given that $\dot{\vec{r}}(t_o) = \vec{\beta}$ and therefore,

$$m\vec{\beta} = 12t_o \cdot \vec{a} + (\sin t_o)\vec{b} + \vec{u}$$

which gives $\vec{u} = m\vec{\beta} - 12t_o \cdot \vec{a} - (\sin t_o)\vec{b}$ and therefore,

$$m\dot{\vec{r}}(t) = 12(t - t_o)\vec{a} + (\sin t - \sin t_o)\vec{b} + m\vec{\beta}.$$

Integrating this equation now, we get

$$m\vec{r}(t) = 6(t - t_o)^2\vec{a} - [\cos t + t\sin t_o]\vec{b} + mt\vec{\beta} + \vec{v}. \qquad (*)$$

where $\vec{v}$ is another constant vector to be determined from the given data: $\vec{r}(t_o) = \vec{\alpha}$. Substituting $t = t_o$ and $\vec{r}(t_o) = \alpha$ in the last equation above, we get

$$m\vec{\alpha} = -(\cos t_o + t_o \sin t_o)\vec{b} + mt_o\vec{\beta} + \vec{v}.$$

and, therefore

$$\vec{v} = m\vec{\alpha} + (\cos t_o + t_o \sin t_o)\vec{b} - mt_o\vec{\beta}.$$

This value of $\vec{v}$ substituted in $(*)$ gives

$$m\vec{r}(t) = 6(t - t_o)^2\vec{a} + [\cos t_o - \cos t + (t_o - t)\sin t_o]\vec{b} + m(t - t_o)\vec{\beta} + m\vec{\alpha}$$

which finally gives the trajectory

$$t \longmapsto \vec{r}(t) = \frac{6}{m}(t - t_o)^2\vec{a} + \frac{1}{m}[\cos t_o - \cos t + (t_o - t)\sin t_o]\vec{b} + (t - t_o)\vec{\beta} + \vec{\alpha}. \qquad \square$$

Example 11 A particle having mass m is acted upon by a force

$$\vec{F}(\vec{r}, \vec{v}, t) = ax\vec{i} + (b\dot{x} + c\dot{y})\vec{j} + (dz + et)\vec{k}$$

where a, b, c, d and e are constants.

At time $t = 0$, the particle is on the positive side of the X-axis and at a distance s from the origin, moving with the speed w along the X-axis. What is the acceleration of the particle at time $t = t_o$?

Solution Initial velocity of the particle is given by

$$\dot{x}(t_o) = w, \dot{y}(t_o) = 0, \dot{z}(t_o) = 0$$

Now, the equation of motion of the particle is

$$m\vec{a}(t) = ax(t)\vec{i} + [b\dot{x}(t) + c\dot{y}(t)]\vec{j} + [d \cdot z(t) + et]\vec{k}.$$

In particular,

$$m\vec{a}(t_o) = ax(t_o)\vec{i} + [b\dot{x}(t_o) + c\dot{y}(t_o)]\vec{j} + [dz(t_o) + et_o]\vec{k}.$$

But it is given that $x(t_o) = s, y(t_o) = 0 = z(t_o), \dot{x}(t_o) = w, \dot{y}(t_o) = 0, \dot{z}(t_o)$.
Therefore, the acceleration $\vec{a}(t_o)$ is given by

$$\vec{a}(t_o) = \frac{1}{m}[as\vec{i} + bw\vec{j} + et_o\vec{k}]$$

which is obtained by substituting the values listed in (∗∗) in the equation (∗).

□

Example 12 The force acting on a particle of mass m is given by

$$F = \vec{F}(\vec{v}) = -a^2\dot{x}\vec{i}$$

a being a constant.

At time $t = t_o$ the particle was at the origin and from there it was set in motion with velocity $u\vec{i}$. Find the position of the particle as a function of time.

Solution The equation of motion of the particle is

$$m\ddot{\vec{r}}(t) = -a^2\dot{x}(t)\vec{i}.$$

Separating the components of the equation, we get

$$\begin{aligned}
m\ddot{x}(t) &= -a^2\dot{x}(t) \\
m\ddot{y}(t) &= 0 \\
m\ddot{z}(t) &= 0.
\end{aligned}$$

The coordinates of the initial position of the particle are

$$x(t_o) = y(t_o) = z(o) = 0,$$

while its initial velocity has the components: $\dot{x}(t_o) = u$, $\dot{y}(t_o) = \dot{z}(t_o) = 0$.

Now integrating $m\ddot{x}(t) = -a^2\dot{x}(t)$ we get $m\dot{x}(t) = -a^2 x(t) + \alpha$ where α is a constant. Putting $t = t_o, x(t_o) = 0, \dot{x}(t_o) = u$ in the above equation, we get $\alpha = mu$ and consequently we have

$$m\dot{x}(t) = -a^2 x(t) + mu.$$

This gives us

$$\frac{dx}{\left(u - \frac{a^2}{m}x\right)} = dt$$

which on integration gives:$-\frac{m}{a^2}\log(u - \frac{a^2}{m}x) = t + \beta$. where β is a constant. Now $x(t_o) = 0$ gives

$$-\frac{m}{a^2}\log u = t_o + \beta.$$

Therefore $-\frac{m}{a^2}\log[u - \frac{a^2}{m}x(t)] = (t - t_o) - \frac{m}{a^2}\log u$ or equivalently, $t = t_o + \frac{m}{a^2}\log\left[\frac{u}{u - \frac{a^2}{m}x(t)}\right]$ which simplifies to

$$\frac{a^2}{m}[t - t_o] = \log \frac{1}{\left[1 - \frac{a^2}{mu}x(t)\right]}$$

and hence

$$x(t) = \frac{mu}{a^2}\left[e^{-\frac{a^2}{m}(t-t_o)}\right]$$

$\ddot{y}(t) \equiv 0 = \dot{y}(t_o) = y(t_o)$ implies $y(t) \equiv 0$ and similarly, $z(t) \equiv 0$.

$\square$

Example 13 A particle is attracted towards a fixed line by a force, the direction of which is perpendicular to the straight line, its magnitude being proportional to the distance of the particle from the line. Show that the orbit of the particle is a curve traced on an elliptical cylinder with the given line as its axis.

Solution We choose the frame $\mathcal{F}$ in such a way that its Z-axis coincides with the given line. Now, it readily follows that the equation of motion of the particle is

$$m\ddot{\vec{r}}(t) = -k\left[x(t)\vec{i} + y(t)\vec{j}\right] \qquad (*)$$

$k > 0$, being the constant of proportionality mentioned in the description of the force field.

Separating the components of the vector equation $(*)$ we get:

$$m\ddot{x}(t) = -kx(t), \quad m\ddot{y}(t) = -ky(t), \text{ and } m\ddot{z}(t) = 0.$$

The solutions of these differential equations can be put in the form

$$\left.\begin{array}{rcl} x(t) & = & A\cos(wt + \alpha) \\ y(t) & = & B\cos(wt + \beta) \\ z(t) & = & Ct + \gamma \end{array}\right\} \qquad (**)$$

where $A, B, C, \alpha, \beta, \gamma$ are some constants and we have written w for $\sqrt{\left(\frac{k}{m}\right)}$ (with an intention to shorten the notation.)

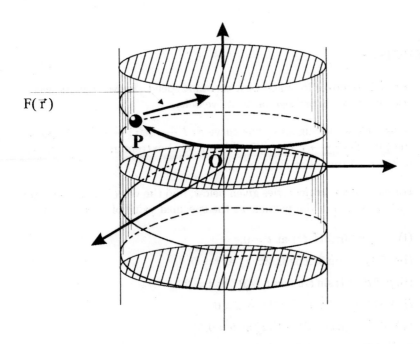

Figure 1.11

Actually, by rotating $\mathcal{F}$ about its Z-axis (if necessary) we can arrange the position of $\mathcal{F}$ in such a way that $\beta = \alpha + \frac{\pi}{2}$ in $(**)$.

We explain this point briefly : First, note that the distance of the particle from the Z-axis of $\mathcal{F}$ remains bounded (in fact, it is $\leq A + B$) . Suppose maximum of the distance is attained at $t = t_o$.

Now we rotate $\mathcal{F}$ and fix its position so that the foot of $\vec{r}(t_o)$ on the XoY-plane falls on the X-axis. Now $x(t_o) = A \cos(wt_o + \alpha) = A$ and $y(t_o) = B \cos(wt_o + \beta) = 0$ hence cos $(wt_o + \alpha) = 1$ while $\sin(wt_o + \beta) = 0$ which gives $wt_o + \alpha + \frac{\pi}{2}) = wt_o + \beta$ that is $\beta = \alpha + \frac{\pi}{2}]$. Thus we assume, $x(t) = A \cos(wt + \alpha)$ and $y(t) = B \cos(wt + \alpha + \frac{\pi}{2}) = B \sin(wt + \alpha)$.

Consequently we have

$$\left(\frac{x(t)}{A}\right)^2 + \left(\frac{y(t)}{B}\right)^2 \equiv \cos(wt + \alpha)^2 + \sin(wt + \alpha)^2 = 1.$$

Thus, the trajectory of the particle

$$t \longmapsto (A\cos(wt+\alpha), B\sin(wt+\alpha), Ct+\gamma)$$

which lies on the ellipse

$$(\frac{x}{A})^2 + (\frac{y}{B})^2 = 1.$$

□

EXERCISES

1. A particle moves in space with constant speed. Prove that its acceleration vector is orthogonal to its velocity.

2. A particle is moving along the curve $c : I \longrightarrow \mathbb{R}^3; t \longmapsto \vec{r}(t)$. Suppose the function $t \longmapsto r(t) = \|\vec{r}(t)\|$ has a local extremum corresponding to $t = t_o$. Prove that $\vec{r}(t_o) \perp \dot{\vec{r}}(t_o)$.

3. For each of the trajectories given below, find the velocity, acceleration and the speed of the particle at the specified value of time.

   (i) $\vec{r}(t) = (6t, 3t^2, t^3)$ at $t = 0$.
   (ii) $\vec{r}(t) = (\sin 3t, \cos 3t, 2t^{3/2})$ at $t = 0$.
   (iii) $\vec{r}(t) = (t\sin t, t \cdot \cos t, \sqrt{3}t)$ at $t = 0$.
   (iv) $\vec{r}(t) = (\sqrt{2} \cdot t, e^t, e^{-t})$ at $t = 0$
   (v) $\vec{r}(t) = (\cos^2 t, 3t - t^3, t)$ at $t = 0$.
   (vi) $\vec{r}(t) = (2\cos t, 3\sin t, t)$ at $t = \pi$.

4. The instantaneous position of a particle is given by the vector $\vec{r}(t)$;

$$\vec{r}(t) = a\left[\cos \ell t \vec{i} + \sin \ell t \vec{j}\right] + \vec{k}$$

   a, b and ℓ being some constants. Find velocity, acceleration of the particle. Also calculate magnitude of each of them.

5. Suppose a particle follows the path $t \longmapsto \vec{r}(t) = (e^t \vec{i} + e^{-t} \vec{j} + \cos t \vec{k})$ untill it flies off along the tangent to the trajectory at the point $\vec{r}(1)$. Find its position vector at $t = 2$.

6. If the particle in Ex. 5 above leaves the trajectory at $\vec{r}(0)$ (instead of $\vec{r}(1)$ to move freely along the tangent, find its position vector at $t = 2$.

7. Suppose that a particle following a curve $c : t \longmapsto \vec{r}(t)$ has the velocity vector $\vec{v}(t)$ given by

$$\vec{v}(t) = (-1, 2, \sin t).$$

Also suppose, $\vec{r}(0) = (2, 2, 4)$. Find the trajectory.

8. A particle following the trajectory $c : t \longmapsto \vec{r}(t)$ has acceleration: $\ddot{r}(t) = (2, 4, e^t)$. Also, suppose its velocity at $t = 0$ is $\vec{v}(0) = (1, 2, 0)$ and it starts from the point $(0, 0, 2)$. Find its trajectory.

9. Suppose velocity of the particle is given by $\vec{v}(t) = \sin t \cdot \vec{i} - \cos t \cdot \vec{j} + 2\vec{k}$. Suppose, $c(0) = \vec{i}$. Describe the curve.

10. A particle of mass 5 units is describing the trajectory $\vec{r}(t) = \sin t \cdot \vec{i} + \cos 2t \cdot \vec{j} + 5t^2 \vec{k}$. Find the force acting on the particle at the times
    (i) $t = 0$ (ii) $t = \frac{\pi}{2}$ (iii) $t = \frac{\pi}{4}$. (iv) $t = 2$.

11. A particle of mass m is acted upon by a force $\vec{F}(t)$ given by

$$\vec{F}(t) = 8 \sin 2t \vec{a} + e^{-t} \vec{b}$$

where $\vec{a}$ and $\vec{b}$ are constant vectors. Find the position vector $\vec{r}(t)$ and the velocity $\vec{v}(t)$, given $\vec{r}(0) = \vec{v}(0) = 0$.

12. A particle of mass m is acted upon two forces. The force $\vec{P}(t)$ is in the direction of the X-axis and the force $\vec{Q}(t)$ is in the XOY-plane along the line $x = y$. Also $P(t)$ ($=$ magnitude of $\vec{P}(t)$) is $= p \cdot \sin kt$ and $Q(t)$ ($=$ magnitude of $\vec{Q}(t)$) is $q \cos kt$; p, q, k being constants. At time $t = 0$ the particle has position $(b, 0, 0)$ and is moving towards the origin with speed $\frac{p}{mk}$. Find the trajectory $t \longmapsto \vec{r}(t)$ of the particle. Prove that the particle moves along an ellipse. Also, find its center and axes.

13. A particle moves under the action of the force $\vec{F}(t) = \cos(qt)\vec{a} - k\vec{r}$, where $\vec{a}$ is a constant vector and q, k are positive constants satisfying $mq^2 \neq k$. When $t = 0$, the particle is at the origin and has velocity $\vec{u}$. Find the trajectory of the particle.

14. A particle is constrained to move along the planar spiral $r = ae^{\theta \cdot \cot(\alpha)}$ in such a way that its radius vector rotates with constant angular speed. Denoting the speed of the particle by $v(t)$, show that the magnitude of the acceleration of the particle is $\frac{v^2}{r}$ and is directed at an angle 2α to the radius vector.

15. A particle moves with a speed u along the curve $r = k(1 + \cos\theta)$. Find

    (a) the polar decomposition of its acceleration
    (b) the magnitude of the acceleration.

(c) the angular speed $\dot{\theta}$.

16. In <u>Exercise 15</u> above, prove that both $\dot{\theta}$ and the magnitude of the acceleration of the particle are proportional to $r^{-1/2}$.

17. A wheel of radius a rolls on a horizontal table with constant angular velocity w about a fixed point 0 on its circumference. A particle moves around the circumference with constant speed v_o relative to the wheel. Prove that the orbit of the particle can be written as

$$r = 2a \cdot \sin \left[\frac{v_o \theta}{v_o + 2aw} \right]$$

(r, θ) denoting the polar coordinates of the particle.

18. A wheel of radius $\underline{a}$ rolls on a horizontal table. The plane of the wheel makes a constant angle α with the horizontal table. The center of the wheel describes a circle of radius 2λ with speed w. Determine the angular velocity of the wheel about its axis.

19. A particle of mass m is located at the end B of a mass-less rod AB having length ℓ. The end A of the rod is free to move along a smooth horizontal wire. The rod is held parallel to the wire and then is released. Determine the horizontal and vertical components of the velocity as functions of the angle made by the rod with the wire.

20. Let $\mathcal{F}$ and $\tilde{\mathcal{F}}$ be two frames of reference having following properties:

(i) Their X-axes coincide.

(ii) $\tilde{\mathcal{F}}$ moves without rotation along the common X-axis with constant speed u relative to $\mathcal{F}$.

Let a particle P be in motion in such a way that with respect to $\mathcal{F}$ it is moving with speed v making a constant angle θ with the X-axis. Prove that its speed v' and the angle θ' with the X-axis in the frame $\tilde{\mathcal{F}}$ are given by

(i) $(v')^2 = u^2 + v^2 - 2u \cdot v \cos \theta$

(ii) $\tan \theta' = \frac{-\sin \theta}{(\cos \theta - \frac{v}{u})}$.

Chapter 2

Rectilinear Motion

*"If I hold the truth in my hand, I should let it go, for, the joy
of pursuing it is greater than that of finding it".*

Sir W. R. Hamilton

2.1 Introduction

In Chapter we studied a few concepts related to the general motion of a
single particle. In this chapter we specialize our study to the *rectilinear
motion* of a particle, that is, the motion of it along a fixed straight line.

Rectilinear motion of a particle is the simplest type of motion. This is
indeed so, mainly for the following two reasons:

(I) The particle is not changing its direction of motion.

(II) The motion can be described in terms of a single Cartesian coordinate-
say, x-coordinate instead of the position vector $\vec{r} = x\vec{i} + y\vec{j} + z\vec{k}$
comprising all the three coordinates (x, y, z).

Apart from being simplest, there is a deeper reason why rectilinear
motion deserves special attention: Often, motion of a complex mechanical
systems has a feature which can be interpreted as the rectilinear motion of
a particle. A separate and detailed study of rectilinear motion of a particle
can naturally give us an insight which will enables us to understand the
complex mechanical system at hand.

Admittedly, this remark is far from self-evident. But instead of elabo-
rating it at this stage, we prefer to illustrate it in terms of some concrete
examples in the chapters to follow.

2.2   Equation of Rectilinear Motion

Suppose, a particle P having mass m is allowed to move only along a straight line L held fix in a force field $\vec{F}$. We choose the frame of reference $\mathcal{F}$ in such a way that its X-axis coincides with the line L. Clearly then, the position vector $\vec{r}(t)$ of the particle is given by

$$\vec{r}(t) = x(t)\vec{i} \simeq x(t) \tag{1}$$

Consequently we have:

$$\left.\begin{array}{rcl} \vec{v}(t) & = & \dot{x}(t)\vec{i} \simeq \dot{x}(t) \\ \vec{p}(t) & = & m\dot{x}(t)\vec{i} \simeq m\dot{x}(t) \\ \vec{a}(t) & = & \ddot{x}(t)\vec{i} \simeq \ddot{x}(t) \end{array}\right\} \tag{2}$$

Thus, many of the vector quantities associated with the motion are expressed in terms of the (scalar) function $t \longmapsto x(t)$ and its derivatives.

We get a similar simplification about the force field acting on the particle. We explain it below:

Let $F : \Omega(\supset L) \longrightarrow \mathbb{R}^3$ be a vector field defined in a region Ω in which the line of motion L is located. At a point $A \left(= x\vec{i} \simeq (x,0,0)\right)$ of L, the force acting on P is $\vec{F}(A) = \vec{F}(x)$. We resolve it into components along and perpendicular to L:

$$\vec{F}(x) = f(x)\vec{i} + \vec{G}(x).$$

where $f(x)\vec{i}$ is the component of $F(\vec{x})$ along L and $G(\vec{x})$, that perpendicular to L.

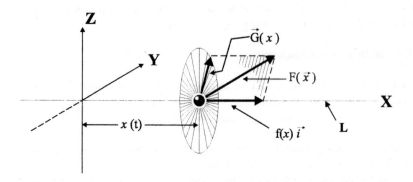

Figure 2.1

Now, since the particle is free to move only along L it is the component $f(x)\vec{i}$ of the applied force $F(\vec{x})$ which is playing the active role in the motion of P . In fact, the other component $G(\vec{x})$ tends to lead P out of L and consequently, there is a reaction to it from the line L (because of the condition that P can move only along L) in the form of the *force of constraint* $\vec{N}(x)$ which just annulls $\vec{G}(x)$. Hence, the total force acting on the particle at a point $A(x,0,0)$ is not $\vec{F}(x)$ but the sum

$$
\begin{aligned}
\vec{F}(x) + \vec{N}(x) &= f(x)\vec{i} + \vec{G}(x) + \vec{N}(x) \\
&= f(x)\vec{i} + \left(\vec{G}(x) + \vec{N}(x)\right) = f(x)\vec{i} + 0 \\
&\quad \text{(since } \vec{G}(x) \text{ and } \vec{N}(x) \text{ are equal and opposite)} \\
&= f(x)\vec{i}
\end{aligned}
$$

Thus, any force field $\vec{F}$ acting in a region in which the line L of motion is located, determines a scalar function $f : \mathbb{R}(\simeq L) \longrightarrow \mathbb{R}$ which is effective in accelerating the particle along L. In other words the function $f : \mathbb{R} \longrightarrow R$ is such that at any point $A(x,0,0)$ of L, the total force acting on P is $f(x)\vec{i}$.

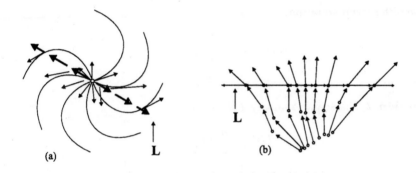

Figure 2.2

The above discussion now makes it clear, that for any force field surrounding L, it is its component field $x \longmapsto f(x)\vec{i}$ along L which we need consider.

Consequently, the equation of motion of P is: $m\ddot{x}\vec{i} = f(x)\vec{i}; x \in \mathbb{R}$ or equivalently, the scalar equation

$$
m\ddot{x} = f(x); x \in \mathbb{R}. \tag{3}
$$

2.3 Integration of Equation of Motion

The simple, scalar form $f : \mathbb{R} \longrightarrow \mathbb{R}$ of the *effective* force field acting on P enables us to associate with the dynamics of P a function $E(x, \dot{x})$ of two variables, - the position x and velocity $\dot{x}$ of the particle. It has the important property of remaining constant along every trajectory of P. In the following, we make use of this property of E to reduce the *second order*

ODE (3) to a *first order* ODE. We integrate this latter equation to get the trajectory $t \longmapsto x(t)$ of P.

We first define a number of terms leading to the definition of the total energy function $E(x, \dot{x})$.

Recall, the (effective) force along L is represented by a function

$$f : \mathbb{R} \longrightarrow \mathbb{R}. \tag{4}$$

Define $U : \mathbb{R} \longrightarrow \mathbb{R}$ by the equation

$$U(x) = - \int_{o}^{x} f(y)dy \tag{5}$$

for all x in $\mathbb{R}$

Definition 1 U is the **potential energy function** of the force field. For any x in $\mathbb{R}$, $U(x)$ is the potential energy of P at the point x.

Note that the potential energy function U determines the force field f by the differential equation

$$f(x) = -\frac{dU}{dx}(x) \tag{6}$$

for all x in $\mathbb{R}$.

Definition 2 The scalar quantity $T(t)$ given by

$$T(t) = \frac{m}{2}\dot{x}(t)^2 \tag{7}$$

is the instantaneous **kinetic energy** of the particle.

Clearly, $T(t)$ depends on the instant t *via* the velocity $\dot{x}(t)$. Often we are interested in its dependence on $\dot{x}(t)$ and **not** on the instant t. Consequently, it is expressed as a function of $\dot{x}$ by suppressing the time t:

$$T = T(\dot{x}) = \frac{m}{2}\dot{x}^2 \tag{8}$$

Definition 3 The sum $T(\dot{x}) + U(x) = \frac{m}{2}\dot{x}^2 + U(x)$ is the **total energy** of the particle. We denote the total energy by $E(x, \dot{x})$. Thus

$$E(x, \dot{x}) = \frac{m}{2}\dot{x}^2 + U(x) \tag{9}$$

Note that $U(x)$ depends on the position coordinate x of P while $T(\dot{x})$ depends on the velocity $\dot{x}$ of it (while at x) and therefore, the sum E is indeed a function of both, x and $\dot{x}$; $E = E(x, \dot{x})$.

But along a particular trajectory of P, both x and $\dot{x}$ are parametrized by the time t. Consequently, we have $E = E\left(x(t), \dot{x}(t)\right) = E(t)$.

We prove below that the function $t \longmapsto E(t)$ is constant and consequently, the function $E(x, \dot{x})$ is constant along each trajectory of the particle.

Proposition 1 The total energy remains constant along each trajectory of the particle.

Proof We consider any trajectory $t \longmapsto x(t)$ and the energy function $t \longmapsto E(x(t), \dot{x}(t))$ along it. We have:

$$
\begin{aligned}
\frac{d}{dt} E(x(t), \dot{x}(t)) &= \frac{d}{dt}\left[\frac{m}{2}\dot{x}(t)^2 + U(x(t))\right] \\
&= m\dot{x}(t)\ddot{x}(t) + \frac{dU}{dx}(x(t))\dot{x}(t) \\
&= \left[m\ddot{x}(t) + \frac{dU}{dx}(x(t))\right]\dot{x}(t) \\
&= 0 \quad \text{using (3) and (6)}
\end{aligned}
$$

Hence $t \longmapsto E(x(t)$ is constant. $\qquad\qquad\qquad\qquad\qquad\square$

In particular, for a conveniently chosen instant t_o we have

$$
E(x(t), \dot{x}(t)) = E(x(t_o), \dot{x}(t_o)) \tag{10}
$$

This result is known as the **principle of conservation of the total energy**.

Consider a particular trajectory and denote by E the *constant value* of the total energy of P along the chosen trajectory. Thus, we have $E = \frac{m}{2}\dot{x}(t)^2 + U(x(t))$ which gives

$$
\dot{x}(t) = \sqrt{\frac{2}{m}[E - U(x(t))]} \tag{11}
$$

On separating the variables, we get $dt = \frac{dx}{\sqrt{\frac{2}{m}[E-U(x)]}}$ which on integration gives: $t = t_o + \int_{x_o}^{x} \frac{dy}{\sqrt{\frac{2}{m}[E-U(y)]}}$.

Thus the above equation gives us time t as a function of the position x. Using the **inverse function theorem** of mathematical analysis, we get the position coordinate x as a function of the time t : $x = x(t)$.

We solve the following example to make the reader familiar with the working of the problems.

Example 1 A particle having mass m is moving along the X-axis under the attractive force $f(x) = -mk^2\left(x + \frac{a^4}{x^3}\right)$ a and k being positive constants. If it starts from $x = a$ with initial velocity 0, show that it reaches the point $x = 0$ after time $t = \frac{\pi}{4a}$.

Solution We start measuring the time from the instant P leaves $x = a$. Clearly, the potential energy U of P is given by

$$U(x) = \frac{mk^2}{2}(x^2 - \frac{a^4}{x^2}).$$

Now, we have $(x(0), \dot{x}(0)) = (a, 0)$ and therefore

$$
\begin{aligned}
E = E\left(x(0), \dot{x}(0)\right) &= \frac{m}{2}\dot{x}(0)^2 + U(x(0)) \\
&= 0 + \frac{mk^2}{2}(a^2 - \frac{a^4}{a^2}) = 0.
\end{aligned}
$$

Therefore, equation (10) gives $\frac{m\dot{x}^2}{2} + \frac{m \cdot k^2}{2x(t)^2}\left[x(t)^4 - a^4\right] = 0$ and hence $x(t)\dot{x}(t) = \pm k\left[a^4 - x(t)^4\right]^{\frac{1}{2}}$. We must take the negative sign as P is moving towards the origin from $x = a > 0$. Substituting $u(t)$ in place of $x(t)^2$, we get $\frac{1}{2}\dot{u}(t) = -k\sqrt{a^4 - u^2(t)}$ or equivalently $\frac{du}{\sqrt{(a^4 - u^2)}} = -2kdt$ and integrating it we get $\sin^{-1}\left(\frac{u}{a^2}\right) = -2kt + B$, B being a constant. Note that $u = a^2$ when $t = 0$ and therefore, $\sin^{-1}\left(\frac{a^2}{a^2}\right) = -2k \cdot 0 + B$ which gives $B = \frac{\pi}{2}$. Consequently, $\sin^{-1}\left(\frac{x(t)}{a}\right)^2 = -2kt + \frac{\pi}{2}$ that is, $\frac{x(t)^2}{a^2} = \sin\left(-kt + \frac{\pi}{2}\right) = \cos(2kt)$ and therefore $x(t) = a\sqrt{\cos 2kt}$.

Let τ be the time taken by the particle to reach the point $x = 0$. Then $x(\tau) = 0$ gives $a\sqrt{\cos 2k\tau} = 0$ which in turn gives $\tau = \frac{k}{4\pi}$. □

Example 2 A particle P having mass m moves along the X-axis under an attractive force $f(x) = \frac{-\mu}{x^2}(\mu > 0$ being a constant) per unit mass. If P starts from rest when $x = a > 0$, show that the displacement function $t \longmapsto x(t)$ satisfies

$$\sqrt{\left(\frac{x(t)}{a}\right)}\sqrt{\left(1 - \frac{x(t)}{a}\right)} + \cos^{-1}\sqrt{\left(\frac{x(t)}{a}\right)} = \sqrt{\left(\frac{2\mu}{a^3}\right)} \cdot t.$$

Proof We assume $t = 0$ when $x = a$. The force on P being $-\frac{\mu m}{x^2}$, the potential energy function U is given by $U(x) = \frac{-\mu m}{x}$.

Initially, the total energy E is $\frac{m}{2}\dot{x}(0)^2 + U(x(0)) = \frac{-\mu m}{a}$ since $\dot{x}(0) = 0$ and $x(0) = a$. Therefore, equation (10) becomes

$$\frac{m}{2}\dot{x}(t)^2 - \frac{\mu m}{x(t)} = -\frac{\mu m}{a}$$

therefore $\dot{x}(t)^2 = 2\mu\left[\frac{1}{x(t)} - \frac{1}{a}\right]$ which reduces to $\dot{x}(t) = \pm\sqrt{\left(\frac{2\mu}{a}\right)(a - x(t))}$. We fix the negative sign because P is moving from $a > 0$ towards 0

$$\sqrt{x(t)} \cdot \dot{x}(t) = -\sqrt{\left(\frac{2\mu}{a}\right)(a - x(t))}. \qquad (*)$$

Putting $u(t) = \sqrt{x(t)}$, the equation (*) becomes

$$2u(t)^2 \dot{u}(t) = -\sqrt{(\frac{2\mu}{a})}\sqrt{(a - u(t)^2)}$$

or equivalently $\frac{2u^2\,du}{\sqrt{a-u^2}} = -\sqrt{(\frac{2\mu}{a})} \cdot dt$ which simplifies to

$$-2\sqrt{(a - u^2)}du + \frac{2a\,du}{\sqrt{a - u^2}} = -\sqrt{(\frac{2\mu}{a})} \cdot dt.$$

On integration, the above equation gives

$$-2\left\{\frac{u}{2}\sqrt{(a - u^2)} + \frac{a}{2}\sin^{-1}\frac{u}{\sqrt{a}}\right\} + 2a\sin^{-1}\frac{u}{\sqrt{a}} = -\sqrt{(\frac{2\mu}{a})} \cdot t + C$$

that is

$$-\sqrt{x(a - x)} + a\sin^{-1}\sqrt{(\frac{x}{a})} = -\sqrt{(\frac{2\mu}{a})} \cdot t + C \qquad (**)$$

Now $x = a$, when $t = 0$ gives $C = a\sin^{-1}1 = \frac{a\pi}{2}$. Substituting this value of C in (**) and simplifying it we get:

$$\sqrt{x}\sqrt{(a - x)} + a\left(\sin^{-1}\sqrt{(\frac{x}{a})} - \frac{\pi}{2}\right) = -\sqrt{(\frac{2\mu}{a})} \cdot t.$$

that is

$$\sqrt{x}\sqrt{a - x} + a\cos^{-1}\sqrt{(\frac{x}{a})} = \sqrt{(\frac{2u}{a})} \cdot t.$$

Dividing the above equation by a we get the required result. $\qquad\square$

2.4   Some Qualitative Analysis

Because the particle does not leave the straight line at any stage of its motion, it must pass through every point x between any two points $x(t_1)$ and $x(t_2)$ on the trajectory. But this property characterizes an interval. Thus, the set of points visited by the particle, that is its orbit, is an interval on the line of motion.

Of course, there is no reason to believe that the trajectory is the whole of the X-axis. One can ask: Which are the points on the X-axis through which the particle can pass?

The particle must have enough (total) energy to reach various points. Therefore, to answer the question, we consider the equation:

$$E = \frac{m}{2}\dot{x}^2 + U(x)$$

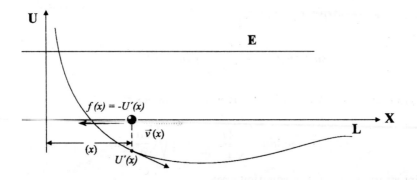

Figure 2.3

Since $\frac{m}{2}\dot{x}^2$ is always non-negative, the inequality $E \geq U(x)$ must be satisfied at every such point through which P passes. Consequently, if x^* is a point on the X-axis with $U(x^*) > E$, then the particle (having total energy E) can never reach it.

Next, suppose P passes through the point x_o at an instant t_o. Suppose, without loss of generality that P moves to the right.

At a time $t \geq t_o$, the velocity of the particle is $\dot{x}(t) = \sqrt{\frac{2}{m}[E - U(x)]}$. Consequently, if $E > U(x)$ for all $x > x_o$, then $\dot{x}(t) > 0$ for all $t > t_o$. This means P does not halt anywhere but goes on moving further to the right and it may even escape to infinity.

On the other hand, there might be a point $\beta > x_o$ with $U(\beta) = E$. For the sake of simplicity, let us assume that the point β has the following properties :

(i) $U(x) < E$ for all x in $[x_o, \beta)$

(ii) $U'(\beta) > 0$ (and so, the graph of U climbs above the line $y \equiv E$).

(iii) (As mentioned above) the point P is moving to the right of x_o.

Then the particle, moving towards β will reach β say at time $t = b > t_o : x(b) = \beta$ and $\dot{x}(b) = 0$. But $m\ddot{x}(b) = -U'(\beta) < 0$. Consequently (at $x = \beta$) through the particle has lost its velocity to move further to the right, it is acted upon by a force $-U(\beta)$ to the left and starts moving backward. Consequently, the point β is called a **turning point** for the particle P (when it is carrying total energy E and moving in the force field f).

The situation is illustrated in Fig. 2.4(a).

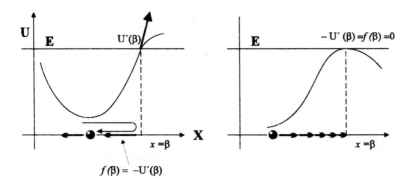

Figure 2.4

If (instead of $U'(\beta) < 0$) we have $U'(\beta) = 0$ in above then it is easy to see that P approaches $x = \beta$ (from $x = x_o$) and once it reaches there, because it has neither velocity nor force $(0 = U'(\beta) = -f'(\beta))$ the particle stays permanently at $x = \beta$. (Fig. 2.4(b)).

Similar things happen when P is moving to the left from. Again there are the following three possibilities:

(I) $U(x) < E$ for all $x \leq x_o$. In this case, leaving x_o, P moves further and further to the left of x_o and it may even escape to $-\infty$. (Fig. 2.5(a)).

(II) There is a point $\alpha < x_o$ having the following properties:

    (i) $U(x) < E$ for all $x \in (\alpha, x_o]$,

    (ii) $U(\alpha) = E$

    (iii) $U'(\alpha) < 0$.

Now, P approaches α from x_o and at α itself, it halts momentarily and then starts in the reverse direction because of the force $U'(\alpha)$ (which is directed to the right.) Thus again α is a **turning point** of P. (Fig. 2.5(b) (III)).

In (II) above, we may have $U'(\alpha) = 0$ instead of $U'(\alpha) < 0$. In this case P moves from x_o to α, but it has neither velocity nor acceleration to move in either direction and consequently, it stays permanently at α. (Fig. 2.5(c)).

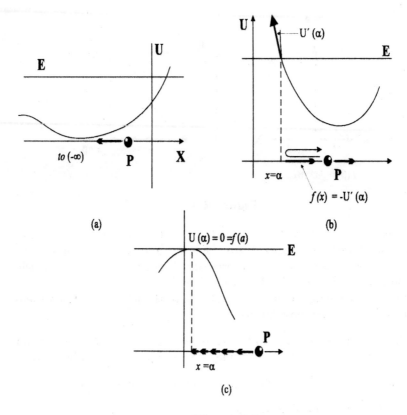

Figure 2.5

Finally, we consider the following situation:

(i) α and β are two consecutive roots of the equation $E = U(x)$ with $\alpha < \beta$.

(ii) $U(x) < E$ for all x in (α, β).

(iii) $U'(\alpha) < 0$ and $U'(\beta) > 0$.

Note that condition (iii) implies that there exists a $\delta > 0$ such that $U(x) > E$ if either $x \in (\alpha - \delta, \alpha)$ or $x \in (\beta, \beta + \delta)$.

The above discussion makes clear that if the particle carrying total energy E is set in motion from anywhere in $[\alpha, \beta]$ then it remains within and performs oscillatory motion between α and β.

Note that for any $x_o \in (\alpha, \beta)$, the velocity is $\pm\sqrt{\frac{2}{m}[E - U(x_o)]}$ and consequently, while passing throw x_o, its speed is the same, irrespective of its direction of motion.

Finally, we obtain an integral expression for the period of the oscillatory motion of P between the turning points α and β. Let us denote it by τ. Note that it is twice the time required by P to travel from α to β.

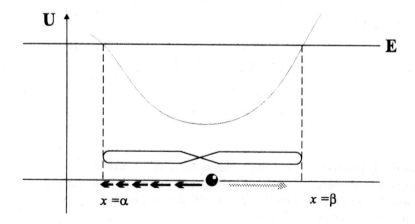

Figure 2.6

Thus

$$\tau = 2 \int_\alpha^\beta dt = 2 \int_\alpha^\beta \frac{dt}{dx} dx = 2 \int_\alpha^\beta \frac{dx}{\dot{x}} = 2 \int_\alpha^\beta \frac{dx}{\sqrt{\frac{2}{m}[E - U(x)]}}$$

$$= \sqrt{(2m)} \int_\alpha^\beta \frac{dx}{\sqrt{[E - U(x)]}} \tag{12}$$

Note that τ depends on the total energy of P.

Example 3 A particle P having mass m is moving in a force field having potential energy function U given by $U(x) = A \mid x \mid^n$ for a natural number $n, A > 0$ being a constant. Find the period of oscillations of P as a function of the total energy $E > 0$.

[Reader should know at least the definition of the Gamma functions to appreciate this example.]

Solution The equation $E = A \mid x \mid^n$ gives the roots $\alpha = -\left(\frac{E}{A}\right)^{\frac{1}{n}}$ and $\beta = \left(\frac{E}{A}\right)^{\frac{1}{n}}$ and therefore, by (12) we have:

$$\tau(E) = \sqrt{2m} \int_{-(E/A)^{\frac{1}{n}}}^{(E/A)^{\frac{1}{n}}} \frac{dx}{\sqrt{[E - A \mid x \mid^n]}}$$

$$= 2\sqrt{2m} \int_0^{(E/A)^{\frac{1}{n}}} \frac{dx}{\sqrt{[E - A \mid x \mid^n]}}$$

$$= 2\sqrt{2m} \frac{1}{\sqrt{E}} \int_0^{(E/A)^{\frac{1}{n}}} \frac{dx}{\sqrt{[1 - \frac{A}{E}x^n]}}.$$

Substituting $\left(\frac{A}{E}\right)^{\frac{1}{n}} x = y$ we get,

$$\tau(E) = 2\sqrt{\left(\frac{2m}{E}\right)} \cdot \left(\frac{E}{A}\right)^{\frac{1}{n}} \int_0^1 \frac{dy}{\sqrt{(1 - y^n)}}$$

$$= \frac{1}{n}\sqrt{\left(\frac{2\pi m}{E}\right)} \left(\frac{E}{A}\right)^{\frac{1}{n}} \frac{\Gamma(\frac{1}{2})}{\Gamma(n + \frac{1}{2})}$$

□

Example 4 A particle of unit mass moves along the X-axis in a force field having potential energy $U(x) = \frac{-x^3}{2}$. If it starts at $t = 0$ from the point 1 with speed 1, obtain the function $t \longmapsto x(t)$ describing the motion. Show that the particle escapes to infinity.

Solution We have

$$E = E(0) = \frac{1}{2}\dot{x}^2(0) - \frac{x^3(0)}{2}$$

$$= \frac{1}{2} \cdot 1 - \frac{1}{2} \cdot 1 = 0.$$

Therefore, by equation (12) we get $\dot{x} = \sqrt{\frac{2}{1}\left[0 + \frac{x^3(t)}{2}\right]} = \pm x^{\frac{3}{2}}(t)$. We take the sign + (Why ?) . This gives $\frac{dx}{x^{\frac{3}{2}}} = dt$ integration of which gives $-2x^{-\frac{1}{2}} = t + C$. Using $x(0) = 1$ we get $C = -2$ and hence $x(t) = \frac{1}{\left(1 - \frac{t}{2}\right)^2}$.

Clearly, the particle escapes to ∞ as $t \longmapsto 2$.   □

Example 5 A particle of mass m is released from rest from a point B which is at a distance $b(> 0)$ from the origin. The attractive force acting on the particle is given by $f(x) = -\frac{k}{x^2}$. Show that the time required for the particle to reach the origin is $\pi \left(\frac{mb^3}{8k}\right)^{\frac{1}{2}}$.

Solution We consider the function U given by $U(x) = \frac{-k}{x}$. Clearly, it is a potential energy function for the given force field.

Now, the total energy E of the particle is

$$E = \frac{m}{2}\dot{x}^2 - \frac{k}{x} \tag{*}$$

Initially $E = -\frac{k}{b}$ (since $x(o) = b, \dot{x}(o) = 0$) therefore, (*) gives

$$-\frac{k}{b} = \frac{m}{2}\dot{x}^2 - \frac{k}{x} \tag{**}$$

On simplification we get

$$\dot{x} = \pm\sqrt{\frac{2k}{m}\left[\frac{1}{x} - \frac{1}{b}\right]}. \tag{***}$$

Since the particle is moving towards the origin, we must consider the negative sign in (∗ ∗ ∗). Rewriting it in the infinitesimal form, we get

$$\frac{\sqrt{x}dx}{\sqrt{(x-b)}} = -\sqrt{\left(\frac{2k}{mb}\right)} \cdot dt.$$

Let τ be the time taken by the particle in traveling from $x = b$ to the origin. Then integration of the above equation gives

$$\int_b^0 \frac{\sqrt{x}dx}{\sqrt{(b-x)}} = -\sqrt{\left(\frac{2k}{mb}\right)} \cdot \tau.$$

Putting $x = u^2$, the above equation becomes

$$\int_{\sqrt{b}}^0 \frac{2u^2 du}{\sqrt{b-u^2}} = -\sqrt{\left(\frac{2k}{mb}\right)} \cdot \tau$$

$$-\frac{\pi b}{2} = -\sqrt{\frac{2k}{mb}} \cdot \tau$$

and therefore $\tau = \pi \sqrt{\left(\frac{mb^3}{8k}\right)}$. □

Example 6 A particle moves in a straight line under an attractive force $f(x) = -\frac{k}{x^{3/2}}$, $k > 0$. Prove that the speed acquired by the particle in moving from ∞ with initial velocity zero to the point $x = a$ equals to that by moving from $x = a$ with starting velocity zero to the point $x = \frac{a}{4}$.

Solution We take $U(x) = \frac{-2k}{x^{1/2}}$ as the potential energy function.

Clearly, the total energy $E(\infty)$ of P at infinity is zero. Let v be the velocity acquired by P in flying from ∞ to a. Then at $x = a$, the total energy of P is $\frac{mv^2}{2} \frac{-2k}{\sqrt{a}}$. Therefore, we get $\frac{mv^2}{2} \frac{-2k}{\sqrt{a}} = 0$ giving

$$v = 2\sqrt{\left(\frac{k}{m}\right)} \cdot a^{\frac{-1}{4}} \qquad (*)$$

Next we consider the second flight of P from $x = a$ to $x = \frac{a}{4}$. It starts with initial velocity 0 and its potential energy $\frac{-2k}{\sqrt{a}}$. Hence its total energy is $\frac{-2k}{\sqrt{a}} = E(a)$.

Let w be the speed gained by the particle when it reaches the point $x = \frac{a}{4}$. Then its total energy is

$$E\left(\frac{a}{4}\right) = \frac{m}{2}w^2 - \frac{2k}{\sqrt{\left(\frac{a}{4}\right)}}.$$

By the conservation of total energy during the second flight, we have

$$\frac{m}{2}w^2 - \frac{2k}{\sqrt{\frac{a}{4}}} = \frac{-2k}{\sqrt{a}}$$

which gives

$$w = 2\sqrt{\left(\frac{k}{m}\right)} a^{\frac{-1}{4}} \qquad (**)$$

Hence $v = w = 2\sqrt{\left(\frac{k}{m}\right)} a^{\frac{-1}{4}}$ $\square$

EXERCISES

1. The velocity of a particle having mass m varies with the displacement according to the equation $v = \frac{b}{x}$, b being a constant. Find the force acting on the particle.

2. A particle having mas m moves along the X-axis, the potential energy function of the force field being given by

$$U(x) = -Ax^\gamma$$

$A > 0, \gamma > 0$ being constants, $\gamma \neq 2$. The particle starts from rest at the point $x = 1$ with momentum $\sqrt{(2mA)}$. Obtain the displacement function $t \longmapsto x(t)$ in terms of m, A and $\beta = 1 - \frac{\gamma}{2}$. For what values of these constants does the motion terminate ? At what time?

3. A particle of mass m moves along the X-axis. The potential U of the force acting on the particle is being given by

$$U(x) = -Ax^4$$

$A > 0$ being a constant. The particle starts at $t = 0$ at the point x_o with momentum $p_o > 0$. Obtain the displacement function $t \mapsto x(t)$.

4. A particle of mass m moves under a force with potential energy $U(x) = \frac{cx}{x^2+a^2}$, a and c being positive constants. If the particle starts from the point $x = -a$ with velocity u find the value of u for which: (a) the motion is oscillatory, (b) the particle escapes to $+\infty$, (c) the particle escapes to $-\infty$.

5. Describe the motion of a particle having mass m in the potential energy function $U(x) = -Ax^4$, $A > 0$ for the case when its energy is zero.

6. Solve for the motion of a particle moving under a force field having potential energy
$$U(x) = -x^{-1} + x^{-2}.$$

Show that for small total energies, the motion is oscillatory but for large energies it is non-periodic and extends to infinity. Find the energy that forms the dividing line between the two cases.

7. Describe the motion of a particle of mass m, where A and α are positive constants:

   (i) $U(x) = \frac{-A}{\cos h^2(\alpha x)}$.

   (ii) $U(x) = -A \tan^2 \alpha x$.

8. A particle P of mass m moves along a straight line through a point 0. Let $x = x(t)$ be the instantaneous position of P relative to 0. When $x > a$, the particle is attracted towards 0 by a force $\frac{mk}{x^2}$ and when $x \leq a$, the particle is repelled by a force $\frac{mak}{x^3}$. If the particle is released from rest at a distance $x = 2a$ from 0, show that it will come to rest instantaneously when $x = \frac{a}{\sqrt{2}}$. Find the time the particle takes to travel from $x = a$ to $x = \frac{a}{\sqrt{2}}$.

9. A particle of mass m is moving along a straight line under the force $f(x) = \frac{-mk^2}{x^3}$, $k > 0$, being a constant. If the particle starts from a point on the straight line at a distance d from 0, with initial velocity zero, prove that it takes time $\frac{d^2}{k}$ to reach the point 0.

10. A and B are two points in a horizontal plane a distance ℓ apart. The points are joined by a light elastic string of natural length ℓ_0 and modulus of elasticity λ with $\ell_o < \ell$. A particle of mass m is attached to the string at its middle point. The particle is drawn aside a small distance perpendicular to AB and released. Show that the particle performs simple harmonic motion and find the period of the simple harmonic motion.

Chapter 3

First Integrals of Motion

"The next great era of awakening of human intellect may well produce a method of understanding the qualitative content of equations".

Richard P. Feynaman

3.1 Introduction

In the first chapter we discussed at length the equation of motion of a particle and studied some of its consequences. In this chapter, we introduce some more notions and their properties related to the motion of the particle.

Recall that the equation of motion involved a number of scalar and vectorial quantities: time, mass, position vector, velocity, acceleration and force. Beside this list, there is a number of auxiliary observable quantities, both scalar and vectorial which also reflect on the motion of the particle. They are total energy, work, linear momentum, angular momentum, torque of the force field about a point etc.

Some of these quantities were introduced to the reader in **Chapter 2** in the context of the rectilinear motion of a particle. It was also proved that the total energy of the particle remains constant along each trajectory of the particle. Subsequently, we made use of the constancy of total energy and reduced the second order ODE of motion to the first order ODE. This was the first hint towards the following two possibilities

(i) There are observable quantities which remain constant along each trajectory of the particle.

(ii) We may use such quantities to simplify the equations of motion (and thereby simplify the whole analysis of the motion). These quantities are called *first integrals* or *constants of motion*. A condition giving rise to a first integral of motion is sometimes called a *conservation principle.*

Besides stating a number of conservation principles, we discuss a few examples to illustrate how first integrals of motion give partial information about the motion and how we can use them to simplify the equation of motion.

As in **Chapter 1** we study the motion of a single particle P having mass m. We choose a stationary frame of reference $\mathcal{F}$ and denote by $\vec{r}(t)$ the instantaneous position vector of P, $\vec{v}(t), \vec{a}(t), \vec{p}(t)$ being respectively its velocity, acceleration and linear momentum, all taken with respect to the frame $\mathcal{F}$.

3.2   Conservative Force Fields

Suppose, the particle is moving in the domain $\Omega(\subset \mathbb{R}^3)$ of the force field F and let $\alpha : J \longrightarrow \Omega$ be any curve. Then the **mechanical work** (or briefly, the *work*) done by the force in the displacement of the particle along the curve α is the line integral

$$\int_\alpha \vec{F}(\vec{r}) \cdot d\vec{r} = \int_a^b \vec{F}(\vec{r}(s)) \cdot \dot{\vec{r}}(s) ds \tag{1}$$

where $J = [a, b]$ and $s \longmapsto \vec{r}(s)$ is a parameterization of α.

We denote the work by $W(\alpha)$.

Before proceeding further, we recall below the following basic properties of the line integrals

(1) The line integral is actually independent of the prametrization, it depends only on the underlying *path* (that is, the set of points sitting on the curve). Thus if $s = s(t)$, $t \in I$ is a reparametrization of the curve then

$$\int_J \vec{F}(\vec{r}(s)) \cdot \dot{\vec{r}}(s) ds = \int_I \vec{F}(\vec{r}(t)) \cdot \dot{\vec{r}}(t) dt.$$

(2) In particular, if the curve $\alpha : J \longrightarrow \Omega$ is taken to be the trajectory of a particle (which is always time parametrized) then we have

$$W(\alpha) = \int_\alpha \vec{F}(\vec{r}) \cdot d\vec{r} = \int_J \vec{F}(\vec{r}(t)) \cdot \dot{\vec{r}}(t) dt = \int_a^b \vec{F}(\vec{r}(t)) \cdot \vec{v}(t) dt$$

(3) If A and B are the end points of a curve $\alpha : J \longrightarrow \Omega$ (that is $J = [a, b]$, then $\alpha(a) = A$ and $\alpha(b) = B$) then $\int_A^B \vec{F}(\vec{r}) \cdot d\vec{r} = - \int_B^A \vec{F}(\vec{r}) d\vec{r}$ both integrals being taken along the same curve α but in the opposite sense.

Example 1 A particle is subjected to the force:

$$\vec{F}(\vec{r}) = 2xz\vec{i} + 3z^2\vec{j} + y^2\vec{k}.$$

Determine the work done by the force in moving the particle along the line $\alpha : x = 2y = 4z$ from the origin to the point $A = (4, 2, 1)$.

Solution Clearly, we can parameterize the curve α using the x-coordinate: $y = \frac{x}{2}, z = \frac{x}{4}$. Then we have

$$\vec{r}(x) = x\vec{i} + \frac{x}{2}\vec{j} + \frac{x}{4}\vec{k}, x \in [0, 4].$$

and Consequently, $d\vec{r}(x) = dx\vec{i} + \frac{dx}{2}\vec{j} + \frac{dx}{4}\vec{k} = \left(\vec{i} + \frac{1}{2}\vec{j} + \frac{1}{4}\vec{k}\right) dx$ and also, $\vec{F}(\vec{r}(x)) = 2x \cdot \frac{x}{4}\vec{i} + 3\left(\frac{x}{4}\right)^2 \vec{j} + \frac{x^2}{4}\vec{k} = \left(\frac{\vec{i}}{2} + \frac{3}{16}\vec{j} + \frac{\vec{k}}{4}\right) x^2$. Consequently,

$$
\begin{aligned}
\vec{F}(\vec{r}) \cdot d\vec{r} &= \left(\frac{\vec{i}}{2} + \frac{3}{16}\vec{j} + \frac{\vec{k}}{4}\right) x^2 \left(\vec{i} + \frac{\vec{j}}{2} + \frac{\vec{k}}{4}\right) dx \\
&= \left(\frac{1}{2} + \frac{3}{32} + \frac{1}{16}\right) x^2 dx = \frac{21}{32} x^2 dx
\end{aligned}
$$

Therefore, the required work is: $W(\alpha) = \frac{21}{32} \int_0^4 x^2 dx = 14$. □

Now, given two points A and B in the domain Ω of the force field, in general, there are several curve in Ω joining A to B. Consider two such curves: $\alpha : I \longrightarrow \Omega$ and $\tilde{\alpha} : \tilde{I} \longrightarrow \Omega$. Then, in general the work $W(\alpha)$ may not be equal to the work $W(\tilde{\alpha})$. But there are force fields $F : \Omega \longrightarrow \mathbb{R}$ for which $W(\alpha)$ and $W(\tilde{\alpha})$ are the same for *any pair of points A, B and for any two curves in Ω joining A to B*. Such force fields are called conservative force fields.

Definition 2 A force field $F : \Omega \longrightarrow \mathbb{R}^3$ is said to be **conservative** if it satisfies the following property: For any curve $\alpha : [a, b] \longrightarrow \Omega, W(\alpha)$ depends only on the end point $\alpha(a), \alpha(b)$ of α and not on α itself.

There is another equivalent formulation of the conservative property $F : \Omega \longrightarrow \mathbb{R}^3$ is conservative if and only if for any **closed curve** $\alpha : [a.b] \longrightarrow \Omega$ (closed in the sense $\alpha(a) = \alpha(b)$) we have $W(\alpha) = 0$.

The reader can (and should) justify the equivalence of the two formulations.

We mention here a criterion for a force field to be conservative $F : \Omega \rightarrow \mathbb{R}^3$ is conservative if and only if it satisfies the condition

$$\nabla \times F = 0 \tag{2}$$

We do not prove it here though we want to use it at a few places in the book. A reader not familiar with the result can find it in a good book on vector analysis.

Assume now that a vector field $F : \Omega \longrightarrow \mathbb{R}^3$ is conservative. Using the conservative property of F, we define a function $U : \Omega \longrightarrow \mathbb{R}$ as follows: Choose any point A_o in Ω as a reference point. Then for any point A in Ω, choose a curve $\alpha : [a, b] \longrightarrow \Omega$ such that $\alpha(a) = A_o$ and $\alpha(b) = A$. Now consider $W(\alpha)$.

According to the assumed conservative property of F, the integral $W(\alpha)$ depends only on A_o and A only. Furthermore, when we are holding A_o fixed and allow A to very in Ω, it will depend only on A giving rise to a well-defined function: $A \longmapsto W(\alpha)$. We put

$$U(A) = -W(\alpha) = - \int_{A_o}^{A} \vec{F}(\vec{r}) \cdot d\vec{r} \qquad (3)$$

Let us note the following two properties of U:

(I) Definition of U clearly depends on the reference point A_o. Replacing A_o by any other point A'_o in Ω, we get

$$- \int_{A'_o}^{A} \vec{F}(\vec{r}) \cdot d\vec{r} = - \int_{A'_o}^{A_o} \vec{F}(\vec{r}) \cdot d\vec{r} - \int_{A_o}^{A} \vec{F}(\vec{r}) \cdot d\vec{r}$$

$$= - \int_{A'_o}^{A_o} \vec{F}(\vec{r}) \cdot d\vec{r} + U(A).$$

showing that the replacement of the point A_o by another point A'_o amounts to adding the constant $- \int_{A'_o}^{A_o} \vec{F}(\vec{r}) \cdot d\vec{r}$ to the function U. Consequently, the function U is not uniquely defined, it *is unique to within an additive constant.*

(II) Function U is related to the force field F by the differential equation

$$\vec{F}(\vec{r}) = - \left[\frac{\partial U}{\partial x}(\vec{r})\vec{i} + \frac{\partial U}{\partial y}(\vec{r})\vec{j} + \frac{\partial U}{\partial z}(\vec{r})\vec{k} \right] \qquad (4)$$

that is $\vec{F}(\vec{r}) = - \operatorname{grad} U(\vec{r})$.

The fact that equation (4) is a consequence of the defining property (3) of the function U is a well-known property discussed in elementary vector analysis.

Definition 3 The function U is called the **potential energy** function of the conservative force field.

Note that, as pointed out earlier, U is not unique (it is unique to within an additive constant and as such the article *the* before the term "potential energy" is really an abuse of languages.

Once a choice of the potential energy function U is made (by fixing the additive constant), the value $U(A)$ of U at a point A of Ω is called the potential energy of the particle at A.

Clearly, the potential energy function U and the work $W(\alpha)$ along a curve α having end points A and B is related by the equation.

$$W(\alpha) = -(U(B) - U(A)) \tag{5}$$

Note that $U(B) - U(A)$ is the change in the potential energy of P as it moves from A to B along α. The negative sign on the right hand side of (5) is the *loss* in potential energy of P during its motion. Consequently, equation (5) expresses the fact that the loss in potential energy incurred by the particle during the motion is utilized in doing the physical work of moving the particle from A to B (along the curve α) in the given force field.

3.3 Energy, Momentum and Torque

We have already considered the *amount of motion* of a moving particle $\vec{p}(t) = m\vec{v}(t)$. We called it the *instantaneous linear momentum* of P.

Definition 4 The quantity $T(t)$ given by $T(t) = \frac{1}{2}m\|\vec{v}(t)\|^2$ is called the instantaneous *kinetic energy* of P.

Sometimes, we suppress the instant t and indicate the dependence of kinetic energy T on the velocity $\vec{v}$ (or on its magnitude v) $T = T(v) = T(\vec{v})$.

Let now $U : \Omega \longrightarrow \mathbb{R}$ denote the potential energy function of the (conservative) force field F.

Definition 5 The sum $E = T + U$ is the **total energy** function of the particle.

Note that E depends on both, the position vector $\vec{r}$ of P (since $U = U(\vec{r})$) and the velocity $\vec{v}$ of the particle (since $T = T(\vec{v})$) $E = E(\vec{r}, \vec{v})$. However, along a particular orbit of the particle, we have the time parameterization $\vec{r} = \vec{r}(t), \vec{v} = \vec{v}(t)$ and consequently $E = E(\vec{r}(t), \vec{v}(t)) = E(t)$.

As in case of rectilinear motion, we prove in the next section that the function $t \longmapsto E(t)$ is constant along *each possible orbit* of P.

Often, we are interested in the rotational aspect of motion of the particle *around a stationary point*. The dynamical observable associated with the rotation of P is its *angular momentum* about the stationary point. We define it below:

Let A be a fix point in space and let $\vec{a}$ denote its position vector.

Definition 6 The **instantaneous angular momentum** of P about A is the vector $(\vec{r}(t) - \vec{a}) \times \vec{p}(t)$. We denote it by $\vec{L}_A(t)$. If the point A (about which the angular momentum is taken) is clear from the context, we may drop the suffix A and write $\vec{L}(t)$ instead of $\vec{L}_A(t)$.

The rotational motion of P about A depends on both (i) the instantaneous position vector $(\vec{r}(t) - \vec{a})$ of P *relative to* A and (ii) the force $\vec{F}(\vec{r})$ acting on P. We designate the vector $\vec{r}(t) - \vec{a} \times \vec{F}(\vec{r})$ to measure the rotational effect of the force. This leads us to the following

Definition 7 (a) The **torque** on P at $\vec{r}$ is the vector $\vec{N}_A(\vec{r}) = (\vec{r} - \vec{a}) \times \vec{F}(\vec{r})$.
(b) The **torque field** of the force field $F : \Omega \to \mathbb{R}^3$ about the point A is
the vector field $N_A : \Omega \longrightarrow \mathbb{R}^3$ given by $\vec{N}_A(\vec{r}) = (\vec{r} - \vec{a}) \times \vec{F}(\vec{r})$, for every
$\vec{r} \in \Omega$.

Again we may drop the suffix A and write $N : \Omega \longrightarrow \mathbb{R}^3$ if the point A
of reference is clear from the context.

Example 2 A particle having mass m moves in a uniform time dependent
force field F given by $\vec{F}(t) = m(t\vec{i} + 2\vec{j} + 4\vec{k})$. If the particle starts from
the origin at time $t = 0$ with initial velocity $2\vec{j} + 2\vec{k}$, calculate the angular
momentum and the torque, both about the origin at $t > 0$.

Solution We have to find $\vec{L} = \vec{L}_o(t) = \vec{r}(t) \times \vec{p}(t) = \vec{r}(t) \times m\vec{v}(t)$ and
$\vec{N}(t) = \vec{N}_o(t) = \vec{r}(t) \times \vec{F}(t)$. Consequently, we must find $t \longmapsto \vec{r}(t)$ and
$t \longmapsto \vec{p}(t)$ first. Now we have $\dot{\vec{p}}(t) = m(t\vec{i} + 2\vec{j} + 4\vec{k})$ by Newton's equation
of motion. Integrating it, we get $\vec{p}(t) = m(\frac{t^2}{2}\vec{i} + 2t\vec{j} + 4t\vec{k}) + \vec{\alpha}$ where $\vec{\alpha}$ is
a constant vector.

In particular, $\vec{p}(0) = \vec{\alpha}$. But it is given that $\dot{\vec{r}}(0) = 2\vec{j} + 2\vec{k}$ and hence
$\vec{p}(0) = m\dot{\vec{r}}(0) = 2m(\vec{j} + \vec{k})(= \vec{\alpha})$.

Therefore, we have $\vec{p}(t) = m \left[\frac{t^2}{2}\vec{i} + 2(t + 1)\vec{j} + 2(2t + 1)\vec{k} \right]$. Also, this
implies $\dot{\vec{r}}(t) = \left[\frac{t^2}{2}\vec{i} + 2(t + 1)\vec{j} + 2(2t + 1)\vec{k} \right]$ integration of which gives
$\vec{r}(t) = \left[\frac{t^3}{6}\vec{i} + (t^2 + 2t)\vec{j} + 2(t^2 + t)\vec{k} \right] + \vec{\beta}$ where $\vec{\beta}$ is a constant vector.
But $\vec{r}(0) = \vec{0}$ (since the particle starts off from the origin) implies $\vec{\beta} = 0$
and hence $\vec{r}(t) = \frac{t^3}{6}\vec{i} + (t^2 + 2t)\vec{j} + 2(t^2 + t)\vec{k}$.

Therefore, the angular momentum $\vec{L}(t)$ is

$$
\begin{aligned}
\vec{L}(t) &= \vec{r}(t) \times \vec{p}(t) \\
&= \left[\frac{t^3}{6}\vec{i} + t(t + 2)\vec{j} + 2t(t + 1)\vec{k} \right] \times m \left[\frac{t^2}{2}\vec{i} + 2(t + 1)\vec{j} + 2(2t + 1)\vec{k} \right] \\
&= m \left[2t^2\vec{i} - \frac{t^3}{3}(t + 2)\vec{j} - \frac{(}{t^3}6(t + 4)\vec{k} \right]
\end{aligned}
$$

Moreover, the torque $\vec{N}(t) = \vec{N}_o(t)$ is given by

$$
\begin{aligned}
\vec{N}(t) &= \vec{r}(t) \times F(t) = \left[\frac{t^3}{6}\vec{i} + (t^2 + 2t)\vec{j} + 2t(t + 1)\vec{k} + 2(t^+t)\vec{k} \right] \\
&\quad \times m \left[t\vec{i} + 2\vec{j} + 4\vec{k} \right] \\
&= m \left[4t\vec{i} - \frac{2}{3}t^2(2t + 3)\vec{j} - \frac{2}{3}t^2(t + 3)\vec{k} \right]
\end{aligned}
$$

$\square$

Recall that in case of the translational motion of the particle, Newton's
equation of motion is a differential equation relating the time derivative of

$\vec{p}(t)$ and the force $\vec{F}(\vec{r}(t))$. Now in the context of the rotational motion of P about A, the counterpart of the pair $\{\vec{p}, F\}$ is the pair $\{\vec{L}_A, N_A\}$. This observation suggests that there should be a differential equation relating $\{\vec{L}_A, N_A\}$.

This is indeed so. In fact, we have the following

Proposition 1 The time rate of change of angular momentum of the particle about a fixed point is equal to the torque of the force about the same point.

Proof: We have

$$
\begin{aligned}
\frac{d}{dt}\vec{L}_A(t) &= \frac{d}{dt}\left[(\vec{r}(t) - \vec{a}) \times \vec{p}(t)\right] \\
&= \dot{\vec{r}}(t) \times \vec{p}(t) + (\vec{r}(t) - \vec{a}) \times \dot{\vec{p}}(t) \\
&= \dot{\vec{r}}(t) \times m\dot{\vec{r}}(t) + (\vec{r}(t) - \vec{a}) \times \vec{F}(\vec{r}(t)) \\
&= 0 + \vec{N}_A(t)
\end{aligned}
$$

Thus we have proved

$$\dot{\vec{L}}_A(t) = \vec{N}_A(t) \tag{6}$$

3.4 Conservation Principles

There are three important conservation principles related to the motion of a particle. They are

(1) Principle of conservation of linear momentum.

(2) Principle of conservation of angular momentum.

(3) Principle of conservation of total energy.

Of the three, we have already proved the first one in **Chapter 1**.

We recall the proof here:

The equation of motion is:

$$\dot{\vec{p}}(t) = F(\vec{r}(t)) \tag{7}$$

If the force acting on the particle is identically zero in the region Ω, then along any trajectory within the region we have $F(\vec{r}(t)) = 0$ and hence $\frac{d}{dt}\vec{p}(t) = 0$ which gives $\vec{p}(t) \equiv \text{constant} = \vec{p}(t_o)$, for any moment t_o during the motion. This proves:

Theorem 1 (Principle Of Conservation Of Linear Momentum.) If the force field acting on the particle is identically zero in a region of the space, then the linear momentum of the particle remains constant while it is passing through the region.

In particular, since the mass of the particle does not change with time, we have $\dot{\vec{p}}(t) = m\dot{\vec{v}}(t)$. Hence $\dot{\vec{p}}(t) \equiv 0$ implies $\dot{\vec{v}}(t) = 0$ and so, the velocity of the particle remains constant. Clearly, the constancy of the velocity *vector* implies that the particle moves along a straight line with constant speed.

Note that (7) is a vector equation and the associated conserved quantity-namely, the linear momentum of the particle - is also a vector quantity.

We now consider the following weaker version of the above conservation principle: Suppose the force field $F : \Omega \longrightarrow \mathbb{R}^3$ acting on P is not identically zero, but has no component in a certain direction. Then we prove below that the component of the linear momentum along that direction remains constant along every trajectory of P while passing through the region Ω.

Theorem 2 Let $\vec{e}$ be a unit vector such that the force field F has no component along $\vec{e}$ throughout the domain Ω. Then along any trajectory, of the particle in Ω, the component of its linear momentum along $\vec{e}$ remains constant.

Proof: We consider the function $t \longmapsto \vec{p}(t) \cdot \vec{e} = p(t)$ along any of the trajectories of P.

We have $\dot{p}(t) = \frac{d}{dt}(\vec{p}(t) \cdot \vec{e}) = \dot{\vec{p}}(t) \cdot \vec{e} = \vec{F}(\vec{r}(t)) \cdot \vec{e} = 0$ for all t with $\vec{r}(t) \in \Omega$; since F has no component along $\vec{e}$. Now $\dot{p}(t) = 0$ implies that the component $p(t)$ of $\vec{p}(t)$ along $\vec{e}$ is constant along any trajectory of P.

This completes the proof of the Theorem.

Next, we consider the equation $\frac{d\vec{L}_A(t)}{dt} = \vec{N}_A(\vec{r}(t))$. Suppose, the force field $F : \Omega \longrightarrow \mathbb{R}^3$ is such that its torque field $N_A : \Omega \longrightarrow \mathbb{R}^3$ taken about a point A is identically zero in $\Omega : N_A \equiv 0$ in Ω. Then we have $\frac{d}{dt}\vec{L}_A(t) \equiv 0$. along any trajectory in Ω. Therefore, the function $t \longmapsto \vec{L}_A(t)$ is constant along the portion of any trajectory in Ω. This proves the following:

Theorem 3 (Principle Of Conservation of the Angular Momentum): If the torque N_A of a force field F taken about a point A is identically zero in a region Ω, then the angular momentum $\vec{L}_A$ remains constant along the portion in Ω of any trajectory of the particle.

Again, we have the following weaker version of the above principle:

Theorem 4 If the torque field N_A has no component in a direction $\vec{e}$, then the component of $\vec{L}_A$ in the direction of $\vec{e}$ remains constant along every trajectory of the particle.

Proof We consider a trajectory of the particle. (Actually, its portion within Ω). We have to prove that the function $t \longmapsto \vec{L}_A(t) \cdot \vec{e}$ is constant on the portion of the trajectory. Now, we have

$$\frac{d}{dt}(\vec{L}_A(t) \cdot \vec{e}) = \left(\frac{d}{dt}\vec{L}_A(t)\right) \cdot \vec{e} = \vec{N}_A(\vec{r}(t)) \cdot \vec{e} \equiv 0.$$

Since N_A has zero component along $\vec{e}$. This proves the constancy of the function $t \longmapsto \vec{L}_A(t) \cdot \vec{e}$.

Actually, rather than the main conservation principles (described in **Theorem 1** and **Theorem 3**,) the weaker versions (**Theorem 2** and **Theorem 4**) have wider applicability.

Finally, we consider the total energy of the particle.

Let $U : \Omega \longrightarrow \mathbb{R}$ be the potential energy function. Recall, the total energy (along an orbit) is $E(t) = \frac{1}{2}m\|\vec{v}(t)\|^2 + U(\vec{r}(t))$. Differentiating it with respect to time t, we get

$$
\begin{aligned}
\frac{d}{dt}E(t) &= \frac{d}{dt}\left(\frac{m}{2}\|\vec{v}(t)\|^2\right) + \frac{d}{dt}U(\vec{r}(t)). \\
&= m\vec{v}(t) \cdot \dot{\vec{v}}(t) + \operatorname{grad} U(\vec{r}(t)) \cdot \dot{\vec{r}}(t). \\
&= \left[m\dot{\vec{v}}(t) + \operatorname{grad} U(\vec{r}(t))\right] \cdot \vec{v}(t) \\
&\qquad \text{by equation } m\dot{\vec{v}}(t) = -\operatorname{grad} U(\vec{r}(t)) \\
&= 0 \cdot \vec{v}(t) = 0.
\end{aligned}
$$

Therefore, the energy function $t \longmapsto E(t)$ is constant along any trajectory of the particle. Thus, we have proved:

Theorem 5 (Principle of Conservation Of Total Energy). The total energy of a particle moving in a conservative force field remains constant along each of its trajectories.

3.5 Use of Conservation Principles

We conclude this chapter by discussing a number of examples which are solved by using the first integrals.

Example 3 A ball B is released from a height b above the ground. It comes down, hits the ground, bounces upwards, comes down again, hits the ground, bounces upwards, $\cdots$. This motion of the ball of comming down and bouncing up is repeated indefinitely. At each impact, the ball looses some of its energy due to the inelastic collision with the ground. Consequently, the speed of the ball after each impact is reduced. It is found that the speed immediately after the impact bears a constant ration with the speed just before impact. This constant ratio-called the *coefficient of restitution*-is denoted by $e(0 < e < 1)$. Assuming that the total time of impact is negligible, prove that the (limiting) time T required for the ball to come to rest is given by:

$$
T = \sqrt{\left(\frac{2g}{b}\right)} \cdot \left[\frac{1+e}{1-e}\right].
$$

Solution We consider the following quantities

(i) The times of successive descents denoted by $T_1, T_2, T_3 \cdots$.

(ii) The speeds of the ball v_i and w_i, just before and just after the ith impact $w_i = ev_i$.

(iii) The maximum height b_i attained by the ball after i^{th} impact.

These quantities are indicated in Fig. 3.1.

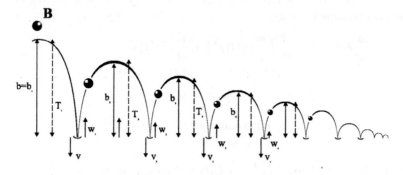

Figure 3.1

Using the principle of conservation of total energy, elementary calculations give us

(1) $v_1 = \sqrt{2gb}, v_2 = ev_1 = e\sqrt{2gb},$
$v_3 = ev_2 = e^2\sqrt{2gb} \cdots v_k = e^{k-1}\sqrt{2gb} \cdots.$

(2) $b_0 = b, b_1 = \frac{v_2^2}{2g} = e^2 b, b_2 = \frac{v_2^2}{2g} = e^4 b \cdots b_k = v^{2k} b = \cdots.$

(3) $T_k = \sqrt{\left(\frac{2b_{k-1}}{g}\right)} = e^{k-1}\sqrt{\left(\frac{2b}{g}\right)}$

The required time T is given by

$$
\begin{aligned}
T &= T_1 + 2T_2 + 2T_3 + \cdots \\
&= \sqrt{\left(\frac{2b}{g}\right)} + 2e\sqrt{\left(\frac{2b}{g}\right)} + 2e^2\sqrt{\left(\frac{2b}{g}\right)} + 2e^3\sqrt{\left(\frac{2b}{g}\right)} + \cdots \\
&= \sqrt{\left(\frac{2b}{g}\right)}\{1 + 2e + 2e^2 + 2e^3 + \cdots\} \\
&= \sqrt{\left(\frac{2b}{g}\right)}\{2(1 + e + e^2 + e^3 + \cdots) - 1\} \\
&= \sqrt{\left(\frac{2b}{g}\right)}\left\{\frac{1+e}{1-e}\right\} = \sqrt{\left(\frac{2b}{g}\right)}\left\{\frac{1+e}{1-e}\right\}.
\end{aligned}
$$

□

Example 4 A particle having unit mass is performing motion in a plane. Its equation of motion is

$$\ddot{\vec{r}}(t) = \frac{-\lambda^2}{r^5}\vec{e_r}(t).$$

It starts to move from a point where $r = a$ with speed $\frac{\lambda}{\sqrt{2}\cdot a^2}$ in any direction. Show that

(1) $h = r^2\dot{\theta}$ remains constant.

(2) $\frac{1}{2}\left[\frac{h^2}{r^4}\left(\frac{dr}{d\theta}\right)^2 + \frac{h^2}{r^2}\right] - \frac{\lambda^2}{4r^4} = 0.$

Solution The first part is proved by using the principle of conservation of the angular momentum. We consider the angular momentum $\vec{L}(t)$ and the torque $\vec{N}(t)$ of the force field, both about the origin. We have

$$\begin{aligned}\vec{N}(\vec{r}(t)) &= \vec{r}(t) \times \vec{F}(\vec{r}(t)) = \vec{r}(t) \times \left(\frac{-\lambda^2}{r^5}\vec{e_r}(t)\right) \\ &= \vec{r}(t) \times \left(\frac{-\lambda^2}{r^6}\vec{r}(t)\right) = \frac{-\lambda^2}{r^6}\vec{r}(t) \times \vec{r}(t) = 0.\end{aligned}$$

Thus we get $\frac{d}{dt}\vec{L}(t) = \vec{N}(\vec{r}(t)) \equiv 0$ giving constancy of the map $t \longmapsto \vec{L}(t)$. In particular we have $\vec{L}(t) = \vec{L}(0)$ for all $t > 0$. On the other hand we have

$$\begin{aligned}\vec{L}(t) &= \vec{r}(d) \times \dot{\vec{r}}(t) = r(t)\vec{e_r}(t) \times \left[\dot{r}(t)\vec{e_r}(t) + r(t)\dot{\theta}(t)\vec{e_\theta}(t)\right] \\ &= 0 + r^2(t)\dot{\theta}(t)\vec{e_r}(t) \times \vec{e_\theta}(t) = r^2(t)\dot{\theta}(t)\vec{k}.\end{aligned}$$

Thus $\vec{L}(t) = r^2(t)\dot{\theta}(t)\vec{k}$. Now the constancy of the map $t \longmapsto \vec{L}(t)$ gives the constancy of the map $t \longmapsto r^2(t)\dot{\theta}(t)$. We denote the constant value of the map by h and get $h = r^2(t)\dot{\theta}(t)$. This proves (1). Next, we have

$$\begin{aligned}\text{grad}\left(\frac{-\lambda^2}{4r^4}\right) &= \frac{d}{dr}\left(\frac{-\lambda^2}{4r^4}\right)\cdot \text{grad}\,(r). \\ &= \frac{\lambda^2}{r^5}\vec{e_r} \quad \text{since grad}(r) = \vec{e_r}. \\ &= -\vec{F}(\vec{r}).\end{aligned}$$

This proves that the force field $F; \vec{r} \longmapsto \frac{-\lambda^2}{r^5}\vec{e_r}$ is conservative and $U; \vec{r} \longmapsto \frac{-\lambda^2}{4r^4}$ is a potential energy function for it. Hence by the conservation principle of total energy we have

$$\begin{aligned}E(0) &= E(t) = \frac{1}{2}\|\dot{\vec{r}}(t)\|^2 + U(\vec{r}(t)) \\ &= \frac{1}{2}\|\dot{\vec{r}}(t)\|^2 - \frac{\lambda^2}{4r^4(t)}.\end{aligned}$$

But

$$\vec{r}(t) = \dot{r}(t)\vec{e_r}(t) + r(t)\dot{\theta}(t)\vec{e_\theta}(t) = \frac{dr}{d\theta} \cdot \dot{\theta}\vec{e_r} + r\dot{\theta}\vec{e_\theta}$$

$$= \left[\frac{dr}{d\theta}\vec{e_r} + r\vec{e_\theta}\right]\dot{\theta} = \frac{h}{r(t)^2}\left[\frac{dr}{d\theta} \cdot \vec{e_r} + r\vec{e_\theta}\right]$$

Consequently, $\|\dot{\vec{r}}(t)\|^2 = \frac{h^2}{r(t)^4}\left[\left(\frac{dr}{d\theta}\right)^2 + r^2\right]$. Now, we get

$$E(t) = \frac{h^2}{r^4(t)}\left[\left(\frac{dr}{d\theta}\right)^2 + r^2\right] - \frac{\lambda^2}{4r^4}.$$

But at $t = 0$, we have

$$E(0) = \frac{\|\dot{\vec{r}}(o)\|^2}{2} + U(\vec{r}(0)) = \frac{\lambda^2}{2 \cdot 2a^4} - \frac{\lambda^2}{4a^2} = 0.$$

Therefore, $E(t) = E(0)$ gives

$$\frac{h^2}{x^4}\left(\frac{dr}{d\theta}\right)^2 + \frac{h^2}{r^2} - \frac{\lambda^2}{4r^4} = 0$$

□

Example 5 A smooth wire is in the shape of a circular helix $x = a \cdot \cos\theta$, $y = a \cdot \sin\theta, z = b \cdot \theta$. A small bead slides on the wire. If it starts from a height b with initial velocity zero, write down the equation resulting from the energy conservation principle and use it to obtain the time required to reach the plane $z = 0$.

Solution Let m be the mass of the particle. We take the coordinate θ as shown in the figure 3.2. Now we have

$$x(t) = a\cos\theta(t), \quad y(t) = a\sin\theta(t), \quad z(t) = b\theta(t) \qquad (*)$$

$(x(t), y(t), z(t))$ being the instantaneous Cartesian coordinates of the bead.
Differentiation of the relations $(*)$ with respect to time gives us

$$\dot{x}(t) = -a \cdot \dot{\theta}(t) \cdot \sin\theta(t), \dot{y}(t) = a \cdot \dot{\theta}(t) \cdot \cos\theta(t), \quad \dot{z}(t) = b\dot{\theta}(t).$$

Now, the kinetic energy $T(t)$ is given by

$$T(t) = \frac{m}{2}\|\dot{\vec{r}}(t)\|^2 = \frac{m}{2}\left[a^2 \cdot \sin^2\theta(t) + a^2 \cdot \cos^2\theta(t) + b^2\right]\dot{\theta}(t)^2$$

$$= \frac{m}{2}(a^2 + b^2)\dot{\theta}(t)^2 = \frac{m}{2}(a^2 + b^2)\frac{\dot{z}(t)^2}{b^2}.$$

Clearly, the potential energy $U(\vec{r})$ is mgz. Consequently, the total energy E is now given by

$$E = \frac{m}{2b^2}(a^2 + b^2)\dot{z}^2 + mgz.$$

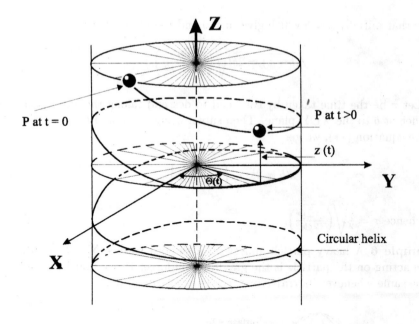

Figure 3.2

Initially, $E = mgb$. Therefore, by the principle of conservation of total energy, we have

$$mgb = \frac{m}{2b^2}(a^2 + b^2)\dot{z}^2 + mgz.$$

which simplifies to $(a^2 + b^2)\dot{z}^2 = 2gb^2(b - z)$ giving $\dot{z}(t) = \frac{\pm b}{\sqrt{(a^2+b^2)}}\sqrt{2g(b - z)}.$

We should consider only the negative sign, because the particle is descending and consequently its z coordinate is decreasing. Therefore we have $\dot{z}(t) = -b\sqrt{\frac{2g(b-z)}{(a^2+b^2)}}.$ which is the required differential equation. We rewrite it in the differential form

$$\frac{dz}{\sqrt{(b - z)}} = -b\sqrt{\left(\frac{2g}{a^2 + b^2}\right)} \cdot dt. \qquad (*)$$

In order to integrate this equation, we put $u = b - z$. Now $du = -dz$ and hence $(*)$ can be rewritten in the form

$$\frac{-du}{\sqrt{u}} = -b\sqrt{\left(\frac{2g}{a^2 + b^2}\right)} \cdot dt$$

Integration of this equation gives us: $\frac{\sqrt{u}}{2} + c = bt\sqrt{\left(\frac{2g}{a^2+b^2}\right)}$; that is

$\frac{1}{2}\sqrt{(b - z)}+c = bt\sqrt{\left(\frac{2g}{a^2+b^2}\right)}$; c being a constant to be determined presently.

Note that initially $z = b$ which gives $c = 0$ and therefore, we get

$$\sqrt{(b-z)} = 2bt\sqrt{\left(\frac{2g}{a^2+b^2}\right)}. \qquad (**)$$

Let τ be the time taken by the bead to descend (along the wire) from the height b to the XOY-plane. Then substituting $z = 0$ and $t = \tau$, in the above equation $(**)$, we get

$$\sqrt{b} = 2b\tau\sqrt{\left(\frac{2g}{a^2+b^2}\right)}.$$

and hence $\tau = \frac{1}{2}\sqrt{\left(\frac{a^2+b^2}{2gb}\right)}.$ □

Example 6 A heavy particle moves on a smooth surface $\sum$. The only force acting on the particle is the gravitational force. Prove that its speed is the same whenever its trajectory crosses a horizontal plane.

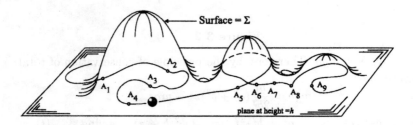

Figure 3.3

Solution: The only force acting on the particle is the constant gravitational force which is conservative. Therefore the total energy of the particle is a first integral of motion (that is it remains constant along the trajectory).

Let $A_1, A_2, A_3, \cdots$ be the points at which the trajectory of the particle crosses the plane.

Let m be the mass of the particle and let $v_1, v_2, v_3, \cdots$ be the speeds of the particle at the points $A_1, A_2, A_3, \cdots$. Then by the conservation principle of total energy, we have

$$E(A_1, v_1) = E(A_2, v_2) = E(A_3, v_3) = \cdots . \qquad (*)$$

where

$$E(A_i, v_i) = \frac{m}{2}v_i^2 + U(A_i), i = 1, 2, 3, \cdots . \qquad (**)$$

Now note that the(gravitational) potential energy is the same at all the points: $U(A_1) = U(A_2) = U(A_3) = \cdots$ because the points are at the same height. Consequently, equations $(*)$ and $(**)$ together imply

$$\frac{m}{2}v_1^2 = \frac{m}{2}v_2^2 = \frac{m}{2}v_3^2 = \cdots .$$

These equalities in turn imply $v_1 = v_2 = v_3 = \cdots$. □

Example 7 A particle having mass m moves under gravity on the inner surface of the paraboloid of revolution $x^2 + y^2 = az$. The particle is initially projected from point $A = (0, 0, a)$ with speed u in the horizontal direction.

(a) Express the speed of the particle in the subsequent motion as a function of the height of the particle above the XOY-plane.

(b) Prove that there are maximum and minimum values of the height of the particle and determine them.

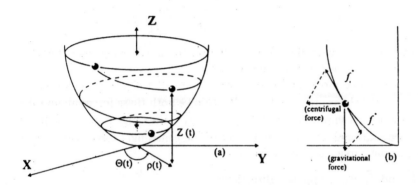

Figure 3.4

Solution We have $E = \frac{m}{2} v^2 + mgz$; E being the (constant) total energy of the particle. Initially, $E = \frac{m}{2} u^2 + mga$. Therefore, we get the equation

$$\frac{m}{2} v^2 + mgz = \frac{m}{2} u^2 + mga.$$

which gives $v = \pm \sqrt{u^2 + 2g(a - z)}$.

We should consider only the positive sign, because $v = +u$ when $z = a$. Thus we get $v = +\sqrt{u^2 + 2g(a - z)}$ which is a function of z.

To calculate the maximum and minimum heights of P, we introduce the cylindrical coordinates (ρ, θ, z) as shown in the figure 3.4. Also, the unit vectors $\vec{e}_\rho(t)$, and $\vec{e}_\theta(t)$ are as shown in the figure 3.4.

Now we have $\vec{r}(t) = \rho(t)\vec{e}_e(t) + z(t)\vec{k}$ and $\dot{\vec{r}}(t) = \dot{\rho}(t)\vec{e}_\rho(t) + \rho(t)\dot{\vec{e}}_\rho(t) + \dot{z}(t)\vec{k} = \rho(t)\vec{e}_e(t) + \rho(t)\dot{\theta}(t)\vec{e}_\theta(t) + \dot{z}(t)\vec{k}$. [Above, we have used the result $\dot{\vec{e}}_\rho(t) = \dot{\theta}(t)\vec{e}_\theta(t)$ which can be proved exactly similarly as in case of the polar coordinates (r, θ)].

Note that the torque (about the origin) of the gravitational force acting on the particle is $\vec{r}(t) \times (-mg\vec{k})$. Clearly it has no component along the z-axis. Consequently, the z-component of the angular moment $\vec{L}(t) = \vec{L}_o(t)$

is a constant scalar. But the angular momentum $\vec{L}(t)$ is given by

$$\vec{L}(t) = m\left[\rho(t)\vec{e}_\rho(t) + z(t)\vec{k}\right] \times \left[\dot{\rho}(t)\vec{e}_e(t) + \rho(t)\dot{\theta}(t)\vec{e}_\theta(t) + \dot{z}(t)\vec{k}\right].$$

$$= (*)\vec{e}_\rho(t) + (**)\vec{e}_\theta(t) + m\rho^2(t)\dot{\theta}(t)\vec{k}.$$

[Here the coefficients of $\vec{e}_\rho(t)$ and $\vec{e}_\theta(t)$ are not need and so we leave the brackets $(*)$ and $(**)$ not calculated]. The z-component of $\vec{L}(t)$ is thus $m\rho^2(t)\dot{\theta}(t)$ which is constant, (as already noted). Initially it was mau. Therefore, we have $mau = m\rho^2(t)\dot{\theta}(t)$ that is $\rho^2(t)\dot{\theta}(t) = au$ giving $\dot{\theta}(t) = \frac{au}{\rho^2(t)} = \frac{u}{z(t)}$. Also the centrifugal force on the particle is

$$m\rho(t)\dot{\theta}^2(t) = m\sqrt{az} \cdot \frac{u^2}{z^2(t)} = \frac{mu^2\sqrt{a}}{z(t)^{3/2}}.$$

Note that the centrifugal force acting on the particle becomes infinite as $z \to 0$ (that is, as the particle goes down).

There are two forces acting on the particle: (i) the gravitational force $-mg\vec{k}$ and (ii) the centrifugal force. We resolve both these forces along the tangent line T as shown in Fig. 3.4(b). Thus $\vec{f}_1$ is the tangential component of the gravitational force and $\vec{f}_2$ is that of the centrifugal force. Now note the following three elementary facts:

(a) $\vec{f}_1$ and $\vec{f}_2$ are in opposite directions.

(b) Magnitude of $\vec{f}_1$ decreases while that of $\vec{f}_2$ increases as the particle approaches the bottom.

(c) The particle will slide down along the paraboloid so long as the magnitude of $\vec{f}_1$ is larger than that of $\vec{f}_2$.

Therefore, there is a horizontal circle S on the paraboloid, as shown in Fig. 3.4(a) at each point of which $\vec{f}_1$ and $\vec{f}_2$ are equal and opposite. Consequently the particle will not be able to leave the circle and the height of S above the XOY-plane will be the minimum height of the particle.

Obviously, the maximum height of the particle is a.

To calculate the minimum height of the particle, we consider the energy equation

$$\frac{m}{2}\rho^2\dot{\theta}^2 + \frac{m}{2}\dot{z}^2 + mgz = \frac{m}{2}u^2 + mga. \text{ or } az\frac{u^2}{z^2} + \dot{z}^2 + 2gz = u^2 + 2ga.$$

Now, at the extreme value of z (= height of the particle) we must have $\dot{z} = 0$. Therefore, the above equation becomes: $\frac{au^2}{z} + 2gz = u^2 + 2ga$.

This a quadratic equation in z and its root are a and $\frac{u^2}{2g}$. Thus the maximum and minimum heights of the particle are a and $\frac{u^2}{2g}$ respectively.   □

Example 8 A particle P of mass m is moving on a smooth horizontal table. A light string attached to the particle passes through a small hole 0 in the table. Initially, the particle is rotating around the hole with speed v_o in a circle of radius a. The string is pulled through the hole untill the radius of the circle is reduced to $\frac{a}{2}$. How much work is done?

Solution The only force acting on the particle is the centrifugal force arising out of the circular motion of the particle about the hole. Let $v(r)$ be the speed of the particle when it is circulating the hole from a distance $r(\frac{a}{2} \leq r \leq a)$.

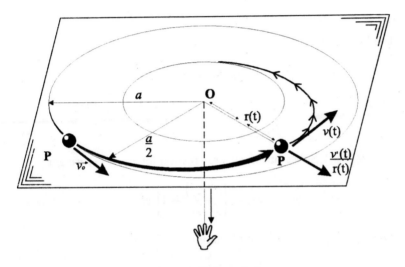

Figure 3.5

We consider the z-component $mrv(r)$ of the angular momentum of particle taken about the hole). Clearly it is a first integral of motion. Initially its value is mav_o. Therefore by the constancy of it, we get the equation $m \cdot r \cdot v(r) = ma \cdot v_o$. This gives us $v(r) = \frac{av_o}{r}$ and hence the centrifugal force acting on the particle when it is at the distance r is $\frac{ma^2v_o^2}{r^3}\vec{e_r}$. Therefore the required work is

$$W = -\int_a^{a/2} \vec{F}(\vec{r}) \cdot d\vec{r} = -\int_a^{a/2} \frac{ma^2v_o^2}{r^3}\vec{e_r} \cdot d\vec{r} = -\int_a^{a/2} \frac{ma^2v_o^2}{r^3}dr = \frac{2mv_o^2}{3}.$$

(We have taken the negative sign because the displacement is in the direction opposite to that of the centrifugal force). ☐

Example 9 A particle P is moves on a surface of revolution under gravity . The equation of the surface is $\sqrt{(x^2 + y^2)} = f(z)$, the z-axis being vertically upwards. If the velocity of the particle at height z_1 is v_1 and that at the height z_2 is v_2, express v_1 and v_2 as functions of z_1 and z_2.

Solution We make use of two conservation principles: (i) Total energy of the particle and (ii) the z-component of the angular momentum of P taken about the origin.

Now, at height z_1 the total energy of the particle is $\frac{m}{2}v_1^2 + mgz_1$ and that at the height z_2 is $\frac{m}{2}v_2^2 + mgz_2$. Therefore by principle (i), we get $\frac{1}{2}m \cdot v_1^2 + mg \cdot z_1 = \frac{1}{2}mv_2^2 + mgz_2$ which simplifies to

$$2g \cdot (z_1 - z_2) = v_2^2 - v_1^2 \tag{$*$}$$

Also, the z component of the angular momentum at height of z_1 is

$$m \cdot \sqrt{(x_1^2 + y_1^2)} \cdot v_1 = mf(z_1) \cdot v_1$$

and at the height of z_2, it is

$$m \cdot \sqrt{(x_2^2 + y_2^2)} \cdot v_2 = m \cdot f(z_2)v_2.$$

Therefore by (ii) above, we get $m \cdot f(z_1) \cdot v_1 = m \cdot f(z_2)v_2$ that is

$$f(z_1) \cdot v_1 = f(z_2) \cdot v_2 \tag{$**$}$$

Hence $v_2 = \frac{f(z_1)}{f(z_2)}v_1$. Substituting this expression for v_2 in $(*)$ we get

$$2g(z_1 - z_2) = \left[\frac{f^2(z_1)}{f(z_2)^2} - 1\right]v_1^2$$

which gives $v_1^2 = \frac{2g(z_1 - z_2)f(z_2)^2}{[f(z_1)^2 - f(z_2)^2]}$ and therefore

$$v_2^2 = \frac{f(z_1)^2}{f(z_2)^2}v_1^2 = \frac{2g(z_1 - z_2)f(z_1)^2}{[f(z_1)^2 - f(z_2)^2]}$$

$\square$

Example 10 One end of an elastic string of unstretched length a and elastic constant λ is attached to a fixed point 0 on a horizontal table. A particle P having mass m is attached to it at the other end. The particle is initially at rest. Then it is given a sudden blow which imparts to it a velocity of magnitude v in a direction normal to the string. In the subsequent motion the string is stretched to a maximum length $3a$. What was the initial speed v of the particle?

Solution Again, we make use of the following two conservation principles: (i) Total energy and (ii) z-component of the angular moment of P taken about the fixed end point of the string.

Now, initially the total energy of the particle is its kinetic energy $\frac{m}{2}v^2$. When the particle is at a distance r from the fixed end of the string, the potential energy of the particle is $U(r) = \frac{\lambda}{2}(r - a)^2$ [since the force on P due to the elastic string is $-\lambda(r - a)$].

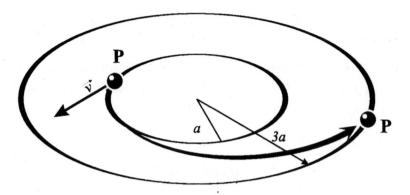

Figure 3.6

Let now the speed of the particle when it is at the distance $3a$ be w. Then the conservation principle (ii) gives us $m \cdot a \cdot v = m \cdot 3a \cdot w$ and therefore $w = \frac{v}{3}$. Consequently, the total energy of P while it reaches the outer circle is: $\frac{m}{2}w^2 + U(3a) = \frac{m}{2}(\frac{v}{3})^2 + 2\lambda a^2$. The conservation of total energy now gives us: $\frac{m}{2}v^2 = \frac{m}{2}(\frac{v}{3})^2 + 2\lambda a^2$. Solving this equation for v, we get $v = 3a\sqrt{(\frac{\lambda}{2m})}$. □

EXERCISES

1. A particle is subjected to the force $\vec{F}(\vec{r}) = 4y\vec{i} + 2x\vec{j} + \vec{k}$. Determine the work done by the force in moving the particle along the helix $x = 4\cos\theta, y = 4\sin\theta, z = 2\theta$ from $\theta = 0$ to $\theta = 2\pi$.

2. A spider having mass m is suspended from the ceiling on an elastic thread of negligible mass having a stretched length β and the natural length α. Calculate the amount of work, the spider is required to do in order to reach the ceiling.

3. A particle is moving in a force field $\vec{F}(\vec{r}) = e^y\vec{i} + (z + xe^y)\vec{j} + (1 + y)\vec{k}$. Calculate the amount of work done by the force in moving the particle along each of the following two paths

   (i) $c_1(t) = t^4\vec{i} + (2t^2 - 1)\vec{j} + (4t^3 - 3)\vec{k}; \ 0 \leq t \leq 1.$

   (ii) $c_2(t) = 2t\vec{i} + \left(\frac{t^2}{2} + \frac{1}{2}\right)\vec{j} + \left(\frac{t^3}{8} + \frac{7}{8}\right)\vec{k}; \ 0 \leq t \leq 1.$

4. Let $F(x,y) = (xy, y^2)$ and let c be the path $y = 2x^2$ joining $(0,1)$ to $(1,2)$ in the plane. Evaluate the work $W(c)$. Does the work depend on the path?

5. Obtain the force field from the potential energy functions:

   (i) $U_1(x,y,z) = \frac{\alpha y}{r^3}.$

   (ii) $U_2(x,y,z) = \frac{\alpha x + \beta y}{r}$ where $r = \sqrt{(x^2 + y^2 + z^2)} > 0.$

6. Find the force fields for which the following are the potential energy functions:

   (i) $U_1(x, y, z) = \frac{1}{2} log \sqrt{(x^2 + y^2 + z^2)}; (x, y, z) \neq (0, 0, 0)$.

   (ii) $U_2(x, y, z) = \frac{1}{2}(k_1 x^2 + k_2 y^2 + k_3 z^2); k_1, k_2, k_3$ being constants.

   (iii) $U_3(x, y, z) = \frac{e^{-kr}}{r}; r = \sqrt{(x^2 + y^2 + z^2)}, (x, y, z) \neq 0$.

7. Show that the force field

$$\vec{F}(\vec{r}) = (y^2 z^2 - 6xyz^3\vec{j} + (3xy^2 z^2 - 6x^2 z)\vec{k}$$

   is conservative and obtain a potential energy function for it.

8. The components of a force field in the XOY-plane are $F_x = A \cdot x, F_y = Bx + Cy^2$. Is this a conservative force field? Calculate the work done by the force field in moving the particle along the close path consisting of the straight lines joining successively the points $(0, 0), (a, 0), (0, b)$, and (a, b).

9. Determine whether each of the following force fields is conservative and if so, find the general potential energy function

   (i) $\vec{F}(\vec{r}) = \vec{a} \times \vec{r}, \vec{a}$ being a non-zero constant vector.

   (ii) $F(\vec{r}) = r\vec{a}, \vec{a}$ being a non-zero constant vector.

   (iii) $F(\vec{r}) = (\vec{a} \cdot \vec{r})\vec{a}, \vec{a}$ being a non-zero constant vector.

10. A force $(-7)\vec{i} + 4\vec{j} + (-5)\vec{k}$ acts at the point $2\vec{i} + 4\vec{j} - 3\vec{k}$. Find

    (i) The torque about the origin exerted by this force.

    (ii) The torque about the point $(1, 1, 1)$ exerted by the same force.

11. A line L passes through the point (a, b, c) and has direction cosines (ℓ, m, n). Determine the component in the direction of L of the torque taken about the point (a, b, c) of the force $\vec{F} = \vec{i}$ acting at the origin.

12. A force with components (A, B, C) acts at the point (a, b, c). Torque of the force is taken about the origin. Determine the component of the torque along the unit vector $\ell\vec{i} + m\vec{j} + n\vec{k}$.

13. A particle of mass five units moves in a uniform time dependent force field

$$\vec{F}(t) = 24t^2\vec{i} + (36t - 16)\vec{j} - 12t\vec{k}.$$

    At time $t = 0$, the particle was located at $3\vec{i} - \vec{j} + 4\vec{k}$ having velocity $6\vec{i} + 15\vec{j} - 8\vec{k}$. Determine the torque of the force and the angular momentum of the particle about the origin for arbitrary $t > 0$.

14. A particle is moving in the XOY-plane under the constant gravitational force along the negative Y-direction. Find four independent first integrals for the two dimensional motion.

15. A particle of mass m is located at the end B of a mass-less rod AB having length a. The end A is free to move on a smooth horizontal wire. The rod is held along the wire and then released. Determine the horizontal and vertical components of the velocity of the particle as functions of the angle that the rod subtends with the wire.

16. A particle of mass m is suspended by means of a mass-less inextensible string of length $3\pi a$ from one end of a horizontal diameter of a cylinder of radius a, lying stationary with its axis horizontal. The particle is set in motion with speed w horizontally so that it moves in a plane perpendicular to the axis of the cylinder and winds the string around it. Prove that the string will not be slack at the top of the path if $w^2 \le 12\pi ga$. If $w^2 = 12\pi ga$, find the tension in the string when the particle is moving vertically upwards for the first time.

17. A toy car moves on a track which is fixed in a vertical plane and contains a circular loop of radius a. Neglecting friction, show that the minimum height above the bottom of the loop from where the car may start from rest and not leave the track at any point is $\frac{5a}{2}$.

18. Consider the motion of a particle which is acted upon by an inverse law of attraction. Writing the equation of motion in the form

$$\ddot{\vec{r}}(t) = -\mu \frac{\vec{r}(t)}{r(t)^3}.$$

show that the vector $\vec{h} = \vec{r}(t) \times \dot{\vec{r}}(t)$ is constant throughout the motion. Also derive the identity $\ddot{\vec{r}}(t) \times \vec{h} = \mu \frac{d}{dt}(\vec{e_r}(t))$ and hence show that the vector $\frac{1}{\mu}(\dot{\vec{r}}(t) \times \vec{h}) - \vec{e_r}(t)$ is constant along the trajectory.

19. A particle is attached to two identical springs of negligible mass, spring constant k and natural length a. One spring is attached at its other end to a fixed point A, the free end of the other spring to a fixed point B such a way that A is at a height $4a$ vertically above B. The particle can move only in the vertical line AB. If $mg < 2ak$, find the position in which the particle can remain at rest and show that in this equilibrium position, both springs are stretched.

20. Consider the particle and springs as described in Exercise 19. Write down the energy equation if the particle is released from rest when the lower spring is at its natural length. Show that the particle has velocity $3a - \left(\frac{mg}{k}\right)$ above B.

Chapter 4

Central Force Fields

"The study of Nature is the most productive source of mathematical discoveries. By offering a specific objective, it provides the advantage of excluding vague problems and unwieldy calculations. It is also a means to formulate mathematical analysis and to isolate the most important aspects to know and to conserve. These fundamental elements are those which appear in all natural effects".

Joseph Fourier

4.1 Introduction

In this chapter, we study the motion of a particle in a force field of a special type. Force fields belonging to this type, called *central force fields* occurs in nature in a number of ways.

As the very name suggests, associated with a central force field is a stationary point O in the physical space V it is the *center* of the force field. The force field is defined in $V \setminus \{O\} = \Omega$. Usually, O is a *singular point* of the force field, that is, the force field can not be extended to include O in its domain. In the domain Ω the force field is characterised by the following property. If A is a point of Ω, then a particle placed at A experiences a force which has direction along OA while its magnitude depends on (that is, is a function of) the distance between O and A.

The gravitational field generated by a stationary massive object around itself is an example of a central force field, the center of the field being the place where the object is situated. Electrostatic field of an electrically charged particle is another prominent example of a central force field.

Central force fields are *conservative* force fields.

73

Another important property of them is the following: The angular momentum $\vec{L}_o(t)$, taken about the force center, of a particle moving in it remains constant along each of the orbits of the particle (that is the angular momentum $\vec{L}_o(t)$ is a first integral of motion of the particle). A consequence of this property is that the particle chooses a plane (passing through the force center) and moves permanently within the plane. Within the plane of motion, the particle either hovers around the force center in a bounded orbit or escapes to infinity. Of course, given any point A in $V \backslash \{O\} = \Omega$ and any vector $\vec{v}_o$, there is a unique trajectory of the particle passing through A with velocity $\vec{v}_o$. Clearly, the plane of motion of this trajectory is the plane determined by the two vectors $\vec{OA}$ and $\vec{v}_o$.

Following few illustrations in Fig. 4.1 give some idea about the complex nature of orbits.

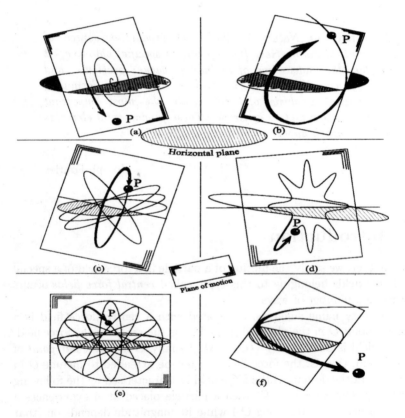

Figure 4.1

Taking into consideration the planar nature of orbit, name, (i) motion is in a plane and (ii) takes place *around* the force center, it is natural to prefer the polar coordinates (r, θ) (in the plane of motion around the force center) to the usual Cartesian coordinates. In the following we will assume (without loss of generality) that the force center is at the origin of $\mathcal{F}$, the motion takes place in the XOX-plane of $\mathcal{F}$ and use the positive X-axis as the initial direction.

First we will spend some time in studying the basic properties of a general central force field and then specialize to the force fields of the type of the inverse square of the distance.

4.2   Basic Properties

Let O be a stationary point in the physical space V.

Definition 1 A **central force field** with center at O is a force field acting on a particle P in the region $\Omega = V \backslash \{O\}$ such that at any point A in Ω, the force acting on P is (i) parallel to $\vec{OA}$ and (ii) has magnitude which is a function of the distance $r (= d(O, A))$ of the point A from O.

Denote the position vector $\vec{OA}$ of A by $\vec{r}$. Writing r for the magnitude of $\vec{r}$, and $\vec{e}_r$ for $\frac{\vec{r}}{r}$ we have the following:

(a) The direction of the force (acting on the particle P while at A), is $\vec{e}_r$.

(b) The magnitude of the force (at A) is a function say $\Phi(r)$ of r. Therefore, the force $F(\vec{r})$ is given by

$$\vec{F}(A) = \vec{F}(\vec{r}) = \Phi(r)\vec{e}_r \tag{1}$$

Thus, the force field $F : \Omega \longrightarrow \mathbb{R}^3$, which is always assumed smooth has the form (1) where $\Phi : (0, \infty) \longrightarrow \mathbb{R}$ is a smooth function. Note that for any point $A(\simeq \vec{r})$ in Ω, if $\Phi(r) > 0$, then the particle is repelled away from O while if $\Phi(r) < 0$, then the particle is attracted towards O.

Suppose, the particle P (executing motion in the force field (1)) has mass m. Then the equation of its motion is:

$$m\ddot{\vec{r}}(t) = \Phi(r(t))\vec{e}_r(t) \tag{2}$$

Now, using the form (1) of the force field, we prove that it is conservative.

Proposition 1 A *central force field* is conservative.

Proof   Let the force field be given by (1). Choose any function $\psi :$ $(0, \infty) \longrightarrow \mathbb{R}$ such that $\frac{d\psi}{dr} \equiv \Phi$. For example, we may take $\psi(r) = \int_1^r \Phi(s)ds$. [Of course we may consider any $a > 0$, in place of 1 and consider $\int_a^r \Phi(s)ds$ instead of $\int_1^r \Phi(s)ds$. But we should avoid $\int_0^r \Phi(s)ds$, it may not exist as Φ may have a singularity at $r = 0$]. Now consider the function given by $U(\vec{r}) = -\psi(\|\vec{r}\|) = -\psi(r)$.

We calculate the gradient of the function U:

$$
\begin{aligned}
\operatorname{grad}_{\vec{r}}(U) &= \operatorname{grad}_{\vec{r}}(-\psi(r)) = -\frac{d\psi}{dr}(r) \cdot \operatorname{grad}(r).\\
&= -\Phi(r)\left[\frac{\partial r}{\partial x}\vec{i} + \frac{\partial r}{\partial y}\vec{j} + \frac{\partial r}{\partial z}\vec{k}\right]\\
&= -\Phi(r)\left[\frac{2x}{2r}\vec{i} + \frac{2y}{2r}\vec{j} + \frac{2z}{2r}\vec{k}\right]r\\
&= \cdot -\Phi(r)\frac{\vec{r}}{r} = -\Phi(r)\vec{e}_r.
\end{aligned}
$$

Thus, we have proved that the function U satisfies the differential equation $\phi(\vec{r})\vec{e}_r = -grad_{\vec{r}}U$ and hence it is a potential energy function for the force field (1). This completes the proof.

Next, we consider the conservation principle of angular momentum $\vec{L}(t) = \vec{L}_o(t)$ of the particle taken about the force center O. Along a trajectory of the particle we have: $\vec{L}(t) = m\vec{r}(t) \times \dot{\vec{r}}(t)$. Differentiating this relation, we get :

$$
\begin{aligned}
\dot{\vec{L}}(t) &= m\dot{\vec{r}}(t) \times \dot{\vec{r}}(t) + m\vec{r}(t) \times \ddot{\vec{r}}(t)\\
&= 0 + \vec{r}(t) \times m\ddot{\vec{r}}(t)\\
&= \vec{r}(t) \times \Phi(r)\vec{e}\vec{r}(t) \ using \ equation \ of \ motion \ (2)\\
&= \Phi(r)\vec{r}(t) \times \frac{\vec{r}(t)}{r(t)} = \frac{\Phi(r)}{r(t)}\vec{r}(t) \times \vec{r}(t) = 0.
\end{aligned}
$$

Hence the equation $t \longmapsto \vec{L}(t)$ remains constant along each trajectory of the particle P. In other words, we have proved the following:

Proposition 2 The angular momentum $\vec{L}(t) = \vec{L}_o(t)$ of the particle (taken about the force center O) remains constant along each trajectory of the particle.

Now, propositions 1 and 2 give rise to two first integrals of motion, one scalar and the other vectorial.**Proposition 1** above and **Theorem 5** of **Chapter 3** imply the conservation of the total energy of the particle: $E = T + U =$ constant or equivalently $\frac{m}{2}\left[\dot{r}^2 + r^2\dot{\theta}^2\right] + U(\vec{r}) =$ constant.

On the other hand, the constancy of the angular momentum function $t \longmapsto \vec{L}(r)$ along a trajectory implies the planar character of motion of the particle. To see this, let us denote the constant angular momentum by $\vec{L}$. Thus $\vec{L}(t) = \vec{L}$ for all t. This equation can be rewritten as

$$
\vec{L} \equiv m\vec{r}(t) \times \dot{\vec{r}}(t)
$$

for all t and therefore $\vec{r}(t)$ remains perpendicular to the constant vector $\vec{L}$. But, vectors perpendicular to a fixed vector remain in a plane through the origin. Thus, the position vectors $\vec{r}(t)$ of the particle remain in the same

plane, namely the plane passing through the origin and perpendicular to
the constant angular momentum vector $\vec{L}$.

To determine the plane of motion, we consider an arbitrarily instant t_o
during the motion and find the position vector $\vec{r}(t_o)$ and the velocity $\vec{v}(t_o)$
of the particle. Now we have $\vec{L} = m\vec{r}(t_o) \times \vec{v}(t_o)$. Clearly the required plane
being the unique plane plane spanned by $\vec{r}(t_o)$ and $\vec{v}(t_o)$.

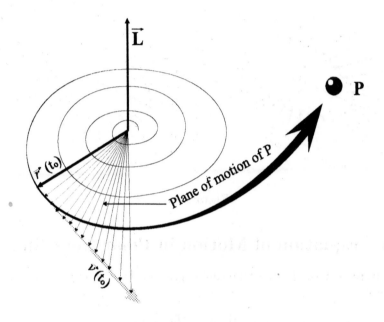

Figure 4.2

Now, we choose a stationary frame of reference $\mathcal{F}$ having the following
properties

(i) Origin of $\mathcal{F} \equiv$ center O of the force field.

(ii) The Z-axis of $\mathcal{F}$ points along the constant vector $\vec{L}$.

Now with this choice of the frame $\mathcal{F}$, it is clear that the plane of the
trajectory of the particle coincides with the XoY-plane of $\mathcal{F}$.

Having fixed the plane of motion of P as the XOY-plane of $\mathcal{F}$ we
consider the circular symmetry of the force field about the force center. If
S is a circle of radius r, center at O, then at all the points $A_1, A_2, A_3 \cdots$ of
S, the forces $\vec{F}(A_1), \vec{F}(A_2), \vec{F}(A_3), \cdots$ have the same magnitude $\Phi(r)$ their
directions being along $OA_1, OA_2, OA_3 \cdots$ either *all* towards O or *all* away
from O.

Under these circumstances, it is certainly convincing that we should
use polar coordinates (r, θ) about the force center in the plane of motion

(that is the XOY-plane of $\mathcal{F}$, instead of the Cartesian coordinates (x, y). Hence, from now onwards, we express the instantaneous dynamical quantities $\vec{r}(t), \vec{v}(t), \vec{a}(t)$ etc. not in terms of the Cartesian coordinates $(x(t), y(t))$ and their time derivatives but in terms of the pair $(r(t), \theta(t))$ and its time-derivatives.

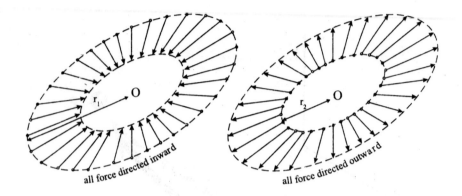

Figure 4.3

4.3   Equation of Motion in Polar Coordinates

In **Chapter 1** we derived following expression for the acceleration of the particle:

$$\ddot{\vec{r}}(t) = \left[\ddot{r}(t) - r(t)\dot{\theta}(t)^2\right] \vec{e}_r(t) + \left[2\dot{r}(t)\dot{\theta}(t) + r(t)\ddot{\theta}(t)\right] \vec{e}_\theta(t).$$

Using it, we rewrite equation (2) of motion in the form

$$m \left\{ \left[\ddot{r}(t - r(t)\dot{\theta}(t)^2\right] \vec{e}_r(t) + \left[2\dot{r}(t)\dot{\theta}(t) + r(t)\ddot{\theta}(t)\right] \right\} = \Phi(r(t))\vec{e}_r(t).$$

Comparing the coefficients of $\vec{e}_r(t)$ and $\vec{e}_\theta(t)$ on both the sides of the above equation, we get

$$
\left.
\begin{array}{lll}
(i) & m[\ddot{r}(t) - r(t)\dot{\theta}(t)^2]. & = \Phi(r(t)) \\
(ii) & m[2\dot{r}(t)\dot{\theta}(t) + r(t)\ddot{\theta}(t)] & = 0
\end{array}
\right\} \qquad (3)
$$

In order to simplify our notations, we do not write the time parameter form now onwards (at least in most of the places). Its presence (or the absence) should be clear to the reader from the context. Now we consider 3(ii) first. Multiplying it by r and integrating the resulting equation with respect to time, we get

$$mr^2\dot{\theta} = \text{ constant.} \qquad (*)$$

To find this constant, we consider the (constant) angular momentum $\vec{L} = L\vec{k}$. We have.

$$\begin{aligned}
\vec{L} &= m\vec{r} \times \dot{\vec{r}} = (mr\vec{e}_r) \times (\dot{r}\vec{e}_r + r\dot{\theta}\vec{e}_\theta) \\
&= mr\dot{r}\vec{e}_r \times \vec{e}_r + mr^2\dot{\theta}\vec{e}_r \times \vec{e}_\theta = 0 + mr^2\dot{\theta}\vec{k}; \quad \text{since } \vec{e}_r \times \vec{e}_\theta = \vec{k}
\end{aligned}$$

Thus, $L\vec{k} = mr^2\dot{\theta}\vec{k}$ giving

$$L = mr^2(t)\dot{\theta}(t). \tag{4}$$

Therefore, the constant appearing in $(*)$ is the magnitude L of the angular momentum of the particle. In particular, we have

$$\frac{1}{2}r^2\dot{\theta} = \frac{L}{2m} = \text{constant.}$$

The expression $\frac{1}{2}r^2\dot{\theta}$ has a geometric significance which we proceed to explain below.

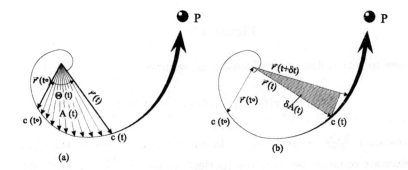

Figure 4.4

We consider a fixed point $c(t_o)$ and a variable point $c(t)$ on the trajectory of the particle. Let $A(t)$ be the area of the sector in the plane of motion bounded by the arms of the vectors $\vec{r}(t), \vec{r}(t_o)$ and the portion of trajectory c lying between the points $c(t_o)$ and $c(t)$. (see Fig. 4.4(a)). Clearly, $A(t)$ is the area swept by the position vector of the particle P as it moves from $c(t_o)$ to $c(t)$ along its trajectory.

Now we prove the following

Proposition 3 $\frac{dA}{dt}(t) = \frac{1}{2}r^2(t)\dot{\theta}(t).$

Proof We consider a third point $c(t + \delta t)$ on the curve which is infinitesimally near to $c(t)$. Let $\delta\theta(t)$ be the angle between the vectors $\vec{r}(t)$ and $\vec{r}(t + \delta t)$. We also consider the area of the infinitesimal sector bounded by the position vectors $\vec{r}(t), \vec{r}(t + \delta t)$ and the portion of c lying between the points $c(t)$ and $c(t + \delta t)$. Let us denote it by $\delta A(t)$. (see Fig.4.4(b)).

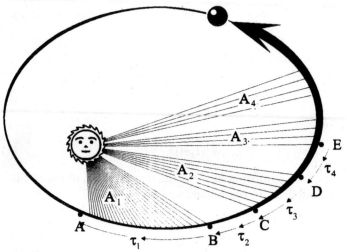

Kepler's second law : (a) Equal areas : $A_1 = A_2 = A_3 = A_4 = \ldots.$

(b) Equal times : $\tau_1 = \tau_2 = \tau_3 = \tau_4 = \ldots.$

Figure 4.5

Then to within first order of smallness, we have

$$\delta A(t) \simeq \frac{1}{2} r(t) \cdot r(t) \cdot \delta\theta(t) = \frac{1}{2} r(t)^2 \delta\theta(t)$$

and therefore $\frac{\delta A(t)}{\delta(t)} \simeq \frac{1}{2} r(t)^2 \frac{\delta\theta(t)}{\delta t}$. In the limit as $\delta t \longrightarrow 0$, the above approximate equation becomes the (perfect) equation: $\frac{dA(t)}{dt} = \frac{1}{2} r(t)^2 \dot\theta(t)$. This completes the proof of the proposition.

Thus, we have proved: A particle moves in a central force field in such a way that the area $A(t)$ swept by the position vector $\vec{r}(t)$ of the particle from an instant t_o to an instant $t(t > t_o)$ has constant time rate of change $\frac{d}{dt} A(t) = \frac{1}{2} \cdot r(t)^2 \cdot \dot\theta(t) = \frac{L}{2m}$.

The same result described in classical terms reads : The motion of the particle in a central force field is such that the position vector of the particle(taken with respect to the force center) *sweeps equal area in equal time*. This is how Johaness Kepler expressed the second of his famous three laws of planetary motion.

Example 1 A particle P having unit mass moves in a plane under the influence of attractive central force field $F = \frac{-2}{r^3} \vec{e_r}$. At time $t = 0$, the distance of the particle from the force center is 2 and the transversal and radial components of its velocity are 1 and $\sqrt{\left(\frac{3}{2}\right)}$ respectively. Show that $\ddot{r}(t) = \frac{2}{r(t)^3}$ and find the function $t \longmapsto r(t)$.

Solution We consider the radial component of the equation of motion

$$\ddot{r}(t) - r(t)\dot{\theta}(t)^2 = -\frac{2}{r(t)^3} \qquad (*)$$

Also, the magnitude L of the initial angular momentum is

$$1 \cdot r^2(0)\dot{\theta}(0) = r(0) \cdot (r(0) \cdot \dot{\theta}(0)) = r(0) \cdot v_\theta(0)$$

where $v_\theta(0)$ is the initial transversal component of the velocity of the particle. It is given that $r(0) = 2$ and $v_\theta(0) = 1$. Therefore, initially, the magnitude of the angular momentum is $1 \cdot 2 \cdot 1 = 2$. Now the expression $L = mr^2\dot{\theta} = r^2\dot{\theta}$ and the conservation principle of angular momentum gives $2 = r^2\dot{\theta}$ or equivalently, $\dot{\theta} = \frac{2}{r^2}$. Putting this expression of $\dot{\theta}$ in $(*)$ we get $\ddot{r} - r(\frac{2}{r^2})^2 = -\frac{2}{r^3}$ which on simplification gives the required result

$$\ddot{r} = \frac{2}{r^3}. \qquad (**)$$

Multiplying the result $(**)$ by $2\dot{r}$ and integrating the new equation with respect to the time t, we get $\dot{r}^2 = -\frac{2}{r^2} + C$ where C is a constant of integration.

Now the data $r(0) = 2$ and $\dot{r}(o) = \sqrt{(\frac{3}{2})}$ gives $C = 2$ and therefore, $\dot{r}(t)^2 = 2 - \frac{2}{r(t)^2}$ which gives $\dot{r}(t) = \sqrt{2}\sqrt{(1 - \frac{1}{r(t)^2})}$ or equivalently $r(t)\dot{r}(t) = \sqrt{2}\sqrt{(r^2(t) - 1)}$.

Putting $r^2(t) = u(t)$, the above equation takes the form

$$\frac{1}{2}\dot{u}(t) = \sqrt{2}\sqrt{(u(t) - 1)}.$$

and therefore $\frac{du}{2\sqrt{(u-1)}} = \sqrt{2}dt$ which, on integration gives

$$\sqrt{u-1} = \sqrt{2} \cdot t + D, \quad D \text{ being a constant.}$$

Since $r^2(0) = u(0) = 4$, we get $D = \sqrt{3}$ therefore, $r(t)^2 - 1 = (\sqrt{2}t + \sqrt{3})^2$ giving $r(t) = \left\{ (\sqrt{2}t + \sqrt{3})^2 + 1 \right\}^{\frac{1}{2}}$. □

Example 2 A particle P having unit mass moves from infinity along a straight line. If continued, it would pass a distance $b\sqrt{2}$ from a point 0. P is attracted towards 0 by a force of magnitude $\frac{k}{r^5}$; $k > 0$ being a constant. The (constant) angular momentum of the particle taken about 0 is $\frac{\sqrt{k}}{b}$. Show that the (r, θ) equation of the orbit is given by $r = b \operatorname{Coth}\left(\frac{\theta}{\sqrt{5}}\right)$.

Solution We have, $\Phi(r)\vec{e_r} = \frac{-k}{r^5}\vec{e_r}$. Clearly, a potential energy function for the force field is $U(r) = \frac{-k}{4r^4}$. Now, $L = 1 \cdot r^2 \cdot \dot{\theta} = r^2\dot{\theta}$. But initially

$L = \frac{\sqrt{k}}{b}$ (given). Hence, by the conservation principle of angular momentum (applied to the magnitude) gives $r^2\dot{\theta} = \frac{\sqrt{k}}{b}$ and hence

$$\dot{\theta} = \frac{\sqrt{k}}{br^2} \tag{*}$$

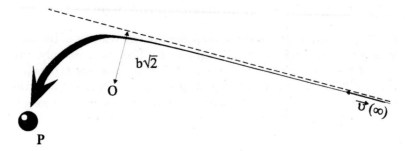

Figure 4.6

Let v_∞ be the initial speed with which the particle was set in motion from infinity. Then the conservation principle of angular momentum gives $L = b\sqrt{2} \cdot v(\infty)$. But it given that $L = \frac{\sqrt{k}}{b}$. Therefore, $b\sqrt{2} \cdot v_\infty = \frac{\sqrt{k}}{b}$ which gives $v_\infty = \frac{1}{b^2}\sqrt{\left(\frac{k}{2}\right)}$.

We calculate the value of the (constant) total energy by calculating it when the particle was set in motion from infinity

$$
\begin{aligned}
E &= E(\infty) = \frac{1}{2} \cdot 1 \cdot v_\infty^2 + U(\infty) \\
&= \frac{k}{4b^4} + 0 = \frac{k}{4b^4}.
\end{aligned}
$$

On the other hand, the general expression for the total energy in terms of the polar coordinates has the form

$$
\begin{aligned}
E &= \frac{1}{2}(\dot{r}^2 + r^2\dot{\theta}^2) - \frac{k}{4r^4} = \frac{\dot{r}^2}{2} + \frac{r^2}{2} \cdot \frac{k}{b^2} \cdot \frac{1}{r^4} - \frac{k}{4r^4} \\
&= \frac{\dot{r}^2}{2} + \frac{k}{2b^2r^2} - \frac{k}{4r^4}.
\end{aligned}
$$

Hence conservation principle of total energy gives the equation

$$\frac{k}{4b^4} = \frac{\dot{r}^2}{2} + \frac{k}{2b^2} \cdot \frac{1}{r^2} - \frac{k}{4r^4}.$$

which simplifies to $2\dot{r}^2 = k\left(\frac{1}{r^4} - \frac{2}{b^2r^2} + \frac{1}{b^4}\right)$ that is,

$$\sqrt{\left(\frac{2}{k}\right)}\,\dot{r} = \pm\left(\frac{1}{r^2} - \frac{1}{b^2}\right) \tag{**}$$

Since at $r = \infty$, we have $\dot{r}(\infty) = v_\infty = \frac{-1}{b^2}\sqrt{\frac{k}{2}}$, (we are attaching the negative sign, because the particle is attracted *towards* 0), we must have the positive sign in (**). Thus we have the equation

$$\sqrt{\left(\frac{2}{k}\right)} \cdot \dot{r} = + \left(\frac{1}{r^2} - \frac{1}{b^2}\right) \qquad (***)$$

Next, we consider $\dot{r}(t)$. Along an orbit, we have

$$\dot{r}(t) = \frac{dr}{d\theta} \cdot \frac{d\theta}{dt} = \dot{\theta}\frac{dr}{d\theta} = \frac{L}{r^2}\frac{dr}{d\theta} = \frac{\sqrt{k}}{b} \cdot \frac{1}{r^2} \cdot \frac{dr}{d\theta}$$

Substituting this expression for $\dot{r}(t)$ in (***) we get $\frac{\sqrt{2}}{b}\frac{1}{r^2}\frac{dr}{d\theta} = (\frac{1}{r^2} - \frac{1}{b^2})$. Writing the same expression in infinitesimal,

$$\frac{dr/r^2}{(\frac{1}{r^2} - \frac{1}{b^2})} = \frac{b}{\sqrt{2}}d\theta.$$

Putting $\frac{1}{r} = u$, using $\frac{-1}{r^2}dr = du$, the above infinitesimal equation takes the form $\frac{du}{(u^2 - \frac{1}{b^2})} = \frac{-b}{\sqrt{2}}d\theta$. We integrate this equation to get

$$\frac{b}{2}\log\left(\frac{\frac{1}{b} + u}{\frac{1}{b} - u}\right) = \frac{b}{\sqrt{2}}\theta + C.$$

C being the constant of integration. Note that at $r = \infty$, $u = 0 = \theta$. Substituting $r = \infty$, $u = 0$ and $\theta = 0$ in the above, we get $C = 0$ and, therefore

$$\log\left(\frac{\frac{1}{b} + \frac{1}{r}}{\frac{1}{b} - \frac{1}{r}}\right) = \sqrt{2} \cdot \theta.$$

or equivalently $\frac{1}{b} + \frac{1}{r} = (\frac{1}{b} - \frac{1}{r})e^{\sqrt{2}\cdot\theta}$. which on simplification gives

$$r = b\left(\frac{1 + e^{\sqrt{2}\cdot\theta}}{e^{\sqrt{2}\cdot\theta} - 1}\right) = b\left(\frac{e^{\frac{\theta}{\sqrt{2}}} + e^{\frac{-\theta}{\sqrt{2}}}}{e^{\frac{\theta}{\sqrt{2}}}} - e^{\frac{-\theta}{\sqrt{2}}}\right) = b \, \coth\left(\frac{\theta}{\sqrt{2}}\right).$$

Thus $r = b \cdot \coth\left(\frac{\theta}{\sqrt{2}}\right)$ is the polar equation of the orbit of P. $\qquad\square$

4.4   Integration of Equations of Motion

We have already seen that the motion of the particle in the given central force field admits two first integrals of motion, namely, (i) the total energy of the particle and (ii) the angular momentum which is a vector quantity. We want to use these constants to simplify the equations of motion of the particle.

We have already used the constant direction of the constant vector $\vec{L}$ in fixing the plane of motion of the particle. Therefore there remain two scalar invariants of motion, the total energy E and the magnitude L of the angular momentum of the particle. Recall $\dot{\theta} = \frac{L}{mr^2}$. Substituting $\frac{L}{mr^2}$ in place of $\dot{\theta}$ in the equation $E = \frac{m}{2}\dot{r}^2 + \frac{m}{2}r^2\dot{\theta}^2 + U(r)$, we get $E = \frac{m}{2}\dot{r}^2 + \frac{L^2}{2m^2r^2} + U(r)$. We put $\frac{L^2}{2mr^2} + U(\vec{r}) = W(\vec{r})$ and call the resulting function $W : \Omega \longrightarrow \mathbb{R}$ the **effective potential energy** function of the particle. Using it we get the energy equation. $E = \frac{m}{2}\dot{r}^2 + W(\vec{r})$.

Using the fact that $W(\vec{r})$ depends on $\vec{r}$ through r only, we will write $W(r)$ instead of $W(\vec{r})$. Hence the energy equation becomes

$$E = \frac{m}{2}\dot{r}^2 + W(r) \tag{5}$$

which gives $\dot{r} = \pm\sqrt{\frac{2}{m}[E - W(r)]}$. The ambiguity in sign arises because there are two possibilities, (a) $t \longmapsto r(t)$ is increasing (b) $t \longmapsto r(t)$ may be decreasing. In a specific problem we are able to decide the sign by taking into consideration the given data of the problem. Usually, the initial conditions and/or the nature of the force field (attractive, repulsive) are used to fix the sign. We separate the variables and integrate the equation

$$\frac{dr}{\pm\sqrt{\frac{2}{m}[E - W(r)]}} = dt \text{ to get } t = \pm \int \frac{dr}{\sqrt{\frac{2}{m}[E - W(r)]}} + C.$$

This equation gives the time t as a function of the distance

$$t = t(r) = \pm \int \frac{dr}{\sqrt{\frac{2}{m}[E - W(r)]}} + C$$

We use the inverse function theorem to get r as a function of $t : r = r(t)$.

Having obtained the distance r as a function of t, we substitute it in the equation $\dot{\theta} = \frac{L}{mr^2}$ to get $d\theta = \frac{L}{mr^2(t)}dt$. Integrating it, we get

$$\theta = \int \frac{L}{mr^2}(t)dt + D.$$

Finally, eliminating t from the functions $r = r(t), \theta = \theta(t)$ we get the equation of the orbit of P in polar coordinates (r, θ).

4.5 An Explanation

Recall, in the **Introduction** section of **Chapter 2**; we stated that in many complex mechanical systems, a particular feature of motion can be interpreted as the rectilinear motion of a (possibly different) particle. We promised to discuss this claim at a latter stage which we do now.

In the present context, the motion of P in the plane has two distinct aspects of motion, namely

(i) The rotational motion around the force center described by the function $t \longmapsto \theta(t)$.

(ii) The variable distance of the P from the force center described by the function $t \longmapsto r(t)$.

All the dynamical equations of the particle (equations of motion in polar coordinates included) involve *both* the functions together with their time derivatives. It was on account of the mixture of the two functions (and their time derivatives) that make the differential equations difficult to solve!

The simplest one (because of its lowest (= first) degree of differentiability) is the energy equation. $E = \frac{m}{2}\left[\dot{r}^2 + r^2\dot{\theta}^2\right] + U(r)$. As noted above, this equation involves not only $r, \dot{r}$ but $\dot{\theta}$ also. But we made use of the constancy of the angular momentum $L = mr^2\dot{\theta}$ to eliminate $\dot{\theta}$ from the equation and got a new equation involving *only* the distance function $t \longmapsto r(t)$: $E = \frac{m}{2}\dot{r}^2 + \frac{L^2}{2mr^2} + U(r)$, that is

$$E = \frac{m}{2}\dot{r}^2 + \frac{L^2}{2mr^2} + U(r). \tag{6}$$

Thus, we have managed to isolate the radial distance function $t \longmapsto r(t)$ in a separate differential equation(6).

Next, recall the energy equation in case of the rectilinear motion for the same particle. (We discussed it in **Chapter 2.**)

$$E = \frac{m}{2}\dot{x}^2 + U(x). \tag{7}$$

Now compare these two equations (6) and (7). The analogy between the two equations sugges to us that the ractalinear distance $x(t)$ of the particle corresponds to the radial distance $r(t)$. Naturally, we expect that the interpreted rectilinear motion should shed light on the actual motion of P.

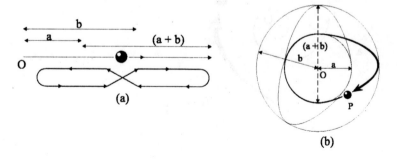

Figure 4.7

Often, this expectation is fulfilled. Here are two illustrative cases:
(I) For example, the oscillatory motion between two points $a, b (a < b)$ on the X-axis corresponds to the bounded motion of P around the force center in the angular region $a \leq r \leq b$. (See fig 4.7) (II) Also, in the straight line motion, if the particle comes from a distance greater than a to $x = a$ from where it is recoiled back, escaping to ∞ afterwards, then the counterpart of this motion in the plane is that the particle comes from outside towards the circle $r = a$, its trajectory is reflected back from the circle and then the particle goes away to infinity from the force center. (See fig 4.8)

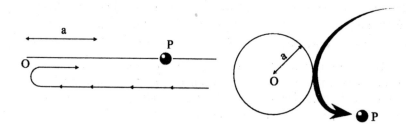

Figure 4.8

Example 3 A particle having mass m is acted upon by an attractive force of magnitude $\frac{m^2}{r^3}$. The particle is projected from a point A which is at a distance a from the force center. The velocity of projection has magnitude $\frac{\sqrt{m}}{a}$ in the direction $\vec{AB}$ where the angle $\angle OAB$ is $\frac{\pi}{4}$. Show that the polar equation of the orbit of the particle is $r = a^{-\theta}$.

Solution Clearly, the polar decomposition of the velocity of projection is $\left(\frac{-\sqrt{m}}{a\sqrt{2}}, \frac{\sqrt{m}}{a\sqrt{2}} \right)$.

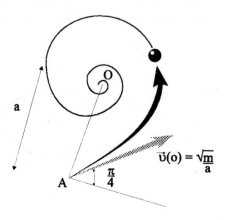

Figure 4.9

The magnitude L of the angular momentum of the particle is $ma \cdot \frac{1}{a}\sqrt{\frac{m}{2}} = m\sqrt{\frac{m}{2}}$. giving rise to the equation $m\sqrt{\left(\frac{m}{2}\right)} = mr^2\dot{\theta}$ which gives

$$\dot{\theta} = \frac{1}{r^2}\sqrt{\left(\frac{m}{2}\right)}. \qquad (*)$$

Also, we have the radial equation of motion $m(\ddot{r} - r\dot{\theta}^2) = -\frac{m^2}{r^3}$ which, on using $(*)$ simplifies to $\ddot{r} - r \cdot \frac{m}{2r^4} = \frac{-m}{r^3}$, that is

$$\ddot{r} = \frac{-m}{2r^3}$$

Multiplying $(**)$ by $2\dot{r}$ and integrating the resulting equation, we get

$$\dot{r}^2 = \frac{m}{2r^2} + C;$$

C being the constant of integration. The initial data $\dot{r} = -\sqrt{\frac{m}{2}} \cdot \frac{1}{a}$ when $r = a$ gives $C = 0$ and so, $\dot{r}(t) = \pm\sqrt{\frac{m}{2}} \cdot \frac{1}{r(t)}$. We should consider the negative sine since initially $\dot{r}$ has negative sign, $\dot{r}(0) = -\sqrt{\left(\frac{m}{2}\right)} \cdot \frac{1}{a}$ and the particle is attracted towards the force center. Thus, $\dot{r}(t) = -\sqrt{\left(\frac{m}{2}\right)} \cdot \frac{1}{r(t)}$.

Also, we have $\dot{r}(t) = \dot{\theta}(t)\frac{dr}{d\theta}(t) = \sqrt{\left(\frac{m}{2}\right)} \frac{1}{r^2(t)} \cdot \frac{dr}{d\theta}$. Therefore, the equation $\dot{r}(t) = -\sqrt{\frac{m}{2}} \frac{1}{r(t)}$ takes the form $\sqrt{\left(\frac{m}{2}\right)} \cdot \frac{1}{r^2}\frac{dr}{d\theta} = -\sqrt{\frac{m}{2}} \frac{1}{r}$ which on simplification becomes $\frac{1}{r}\frac{dr}{d\theta} = -1$. Thus, $\frac{dr}{r} = -d\theta$ which on integration gives $\log r = -\theta + D$. We choose the frame of reference $\mathcal{F}$ in such a way that its X-axis goes along $\vec{0A}$ so that $\theta = 0$ when $r = a$ which gives $D = \log a$ and the equation of the orbit becomes $\log r = -\theta + \log a$, giving $r = ae^{-\theta}$.

□

4.6   Circular Orbits

Suppose, the orbit of the particle is a circle of radius a, its center coinciding with the force center. Thus $r(t) \equiv a$. Differentiating this identity twice, we get

$$(i) \quad \dot{r}(t) \equiv 0, \text{ and } \quad (ii) \quad \ddot{r}(t) \equiv 0 \qquad (8)$$

Equations (5) and 8(i) together imply $E = W(a)$. Hence, the radius a of a circular orbit must be a root of the equation $E = W(r)$. Note that along the circular orbit, $L = ma^2\dot{\theta}$, a being a constant. Thus, the angular velocity of the particle moving along the circular orbit remains constant. Next, equations (3) (i) and (8) (ii) together imply

$$-ma\dot{\theta}^2 = \Phi(a) \qquad (9)$$

But $-ma\dot{\theta}^2 = -\frac{mv^2}{a}$ is the familiar centrifugal force acting on the particle. Thus equation (9) expresses the fact that when the particle is moving along a circular orbit, the central force is equal in magnitude and opposite in direction to the centrifugal force. In other words, when the particle is moving along a circular orbit, the central force balances the centrifugal force at each point of the orbit. Recall that the centrifugal force is always directed away from the force center. (Which is also the center of the circular orbit). Consequently, the balancing act requires that the central force be attractive.

Now, we have $\frac{dW}{dr}(r) = -\left[\Phi(r) + \frac{L^2}{mr^3}\right] = -\left[\Phi(r) + mr\dot{\theta}^2\right]$. Consequently, equation (9) is equivalent to $\frac{dW}{dr}(a) = 0$. Thus we have proved that for the particle to perform circular motion, the following conditions must be satisfied:

$$(i) \quad E = W(a) \quad \text{and} \quad (ii) \quad \frac{dW}{dr}(a) = 0 \qquad (10)$$

a being the radius of the circular motion. Conversely, suppose the two conditions of (10) are satisfied. We prove below that the particle performs circular orbit.

Note that $E = W(a)$ implies that $\dot{r}(t_o) = 0$ for any moment t_o when the particle is at a point on the circle. Now the map $t \longmapsto r(t)$ satisfies the condition (a) $r(t_o) = a$, (b) $\dot{r}(t_o) = 0$ and (iii) $m\ddot{r}(t) = \frac{-dW}{dr}(r(t))$. Clearly $r(t) \equiv a$ is the, solution satisfying the differential equation (iii) together with the initial conditions (i) and (ii).

We have thus proved that a circular orbit of radius $r = a$ can occur if and only if the effective potential energy function W satisfies equation (10).

We, consider the second order derivative $\frac{d^2W}{dr^2}(a)$. We have, either $\frac{d^2W}{dr^2}(a) > 0$ or $\frac{d^2W}{dr^2}(a) \leq 0$. Recall, $\frac{d^2W}{dr^2}(a) > 0$ implies that W has a strict local minimum at $r = a$.

Now, about the orbit. Associated with an orbit is the property of its *stability:* If the orbit is slightly perturbed, (within the same plane) will the new orbit remain close enough to the original orbit ? If the answer is in affirmative, we say that the orbit is **stable.** Otherwise, that is, if a slight perturbation of the orbit results into an orbit which tends to go away from the original orbit, then we say that the orbit is **unstable.** For example, a circular orbit may open and the particle may escape to infinity.

Stability of orbits is an important aspect of dynamics of the whole physical world. For obvious reasons, we want stable orbits. Indeed, the progress of the human race is greatly due to the stability of the Earth. As another example, the orbit of a telecommunication satellite must be stable. Because, if a small change is taking place in the orbit (due to mechanical faults, change in the atmosphere etc.,) then the course of the satellite can be corrected only when the orbit is stable in the above explained sense.

We state a result without proof: A circular orbit corresponding to a local minimum of the function W is a stable circular orbit.

Example 4 A particle P having mass m is moving in the attractive central force field of magnitude $\frac{k}{r^n}$. Prove that if $n < 3$, then the particle can perform a stable circular orbit.

Solution We have $\Phi(r)\vec{e_r}) = \frac{-k}{r^n}\vec{e_r}$ with $k > 0$. A potential energy function for the field is $U(\vec{r}) = \frac{-k}{(n-1)r^{n-1}}$ and when the particle has angular momentum of magnitude L, the effective potential energy function $W(r)$ is given by $W(r) = \frac{L^2}{2mr^2} - \frac{k}{(n-1)r^{n-1}}$.

Therefore, the conditions $\frac{dW}{dr}(a) = 0$, $\frac{d^2W}{dr^2}(a) > 0$ for a stable circular orbit take the form

(i) $\frac{dW}{dr}(a) = \frac{k}{a^n} - \frac{L^2}{ma^3} = 0$ giving $L^2 = \frac{mk}{a^{n-3}}$.

(ii) $\frac{d^2W}{dr^2}(a) = \frac{-kn}{a^{n+1}} + \frac{3L^2}{ma^4} = \frac{-kn}{a^{n+1}} + \frac{3mk}{a^{(n-3)}\cdot ma^4}$ using (1)

$\qquad = \frac{-kn}{a^{n+1}} + \frac{3k}{a^{n+1}} = \frac{(3-n)k}{a^{n+1}}$.

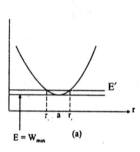

$E = W_{min}$ (a)

(Stable circular orbit corresponding to $E = W_{min}$)

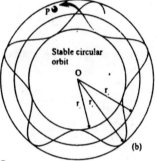

Stable circular orbit

(b)

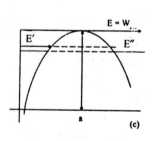

(c)

(Unstable circular orbit corresponding to $E = W_{...}$)

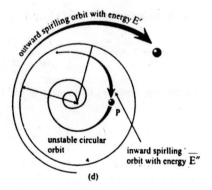

outward spirlling orbit with energy E'

unstable circular orbit

inward spirlling orbit with energy E''

(d)

Figure 4.10

Consequently if $n < 0$, then $\frac{d^2 W}{dr^2}(a) > 0$ giving a local minimum of W at $r = a$, which corresponds to a stable circular. □

4.7   Force Field from Equation of Orbit

Often, the equation of an orbit in the plane of motion is known. We now prove that given the equation of the orbit in the polar coordinates (r, θ), it is possible to obtain the force law $\vec{r} \longmapsto \Phi(r)\vec{e_r}$ along the orbit.

We introduce the variable $u = \frac{1}{r}$. Now, we have

$$
\begin{aligned}
\frac{du}{d\theta} &= \frac{d}{d\theta}(\frac{1}{r}) = \frac{-d}{dr}(\frac{1}{r}) \cdot \frac{dr}{d\theta} = -\frac{1}{r^2}\frac{dr}{d\theta} = -\frac{1}{r^2} \\
&= -\frac{1}{r^2}\frac{\dot{r}}{\dot{\theta}} = -\frac{m\dot{r}}{L}
\end{aligned}
$$

Moreover,

$$
\begin{aligned}
\frac{d^2 u}{d\theta^2} &= \frac{d}{d\theta}\left(\frac{du}{d\theta}\right) = \frac{d}{d\theta}\left(\frac{-m\dot{r}}{L}\right) \\
&= \frac{d}{dt}\left(\frac{-m\dot{r}}{L}\right)\frac{dt}{d\theta} = -\frac{m\ddot{r}}{L\dot{\theta}} = \frac{-m\ddot{r}}{L^2 u^2}.
\end{aligned}
$$

Thus, the equation $\frac{d^2 u}{d\theta^2} = \frac{-m\ddot{r}}{L^2 u^2}$ gives

$$
\ddot{r} = -\frac{L^2 u^2}{m}\frac{d^2 u}{d\theta^2} \tag{$*$}
$$

Also, $r\dot{\theta}^2 = \frac{m^2 r^4 \dot{\theta}^2}{m^2 r^3} = \frac{L^2}{m^2 r^3} = \frac{L^2}{m^2}u^3$, that is

$$
r\dot{\theta}^2 = \frac{L^2 u^3}{m^2} \tag{$**$}
$$

Combining the results: $(*)$ and $(**)$, we get

$$
m(\ddot{r} - r\dot{\theta}^2) = m\left[\frac{-L^2 u^2}{m^2}\frac{d^2 u}{d\theta^2} - \frac{L^2 u^3}{m^2}\right] = -\frac{L^2 u^2}{m}\left[\frac{d^2 u}{d\theta^2} + u\right].
$$

Therefore, the equation $m(\ddot{r} - r\dot{\theta}^2) = \Phi(r)$ takes the form

$$
\Phi(r) = -\frac{L^2}{mr^2}\left[\frac{d^2}{d\theta^2}(\frac{1}{r}) + \frac{1}{r}\right] \tag{11}
$$

The following two examples illustrate how we can use the derivation (11) to get the force law from the given polar equation of the orbit.

Example 5 Find the force field which makes a particle P of mass m, move along the spiral orbit $r = ke^{\alpha\theta}$, k and α being constants.

Solution Given $r = ke^{\alpha\theta}$, we have $u = \frac{1}{r} = \frac{e^{-\alpha\theta}}{k}$. Therefore,

$$\frac{d}{d\theta}\left(\frac{1}{r}\right) = \frac{-\alpha e^{-\alpha\theta}}{k}, \quad \text{and} \quad \frac{d^2}{d\theta^2}\left(\frac{1}{r}\right) = \frac{\alpha^2 e^{-\alpha\theta}}{k} = \frac{\alpha^2}{r}.$$

Now, equation (11) gives us

$$\begin{aligned}
\Phi(r) &= -\frac{L^2}{mr^2}\left[\frac{d^2}{d\theta^2}\left(\frac{1}{r}\right) + \frac{1}{r}\right] \\
&= -\frac{L^2}{mr^2}\left[\frac{\alpha^2}{r} + \frac{1}{r}\right] = -\frac{L^2(\alpha^2 + 1)}{mr^3}.
\end{aligned}$$

Thus, the force field is attractive and varies as the inverse cube of the distance r. □

Example 6 A particle P having mass m moves under the influence of the central force field $-\frac{k}{r^n}\vec{e_r}$. If the orbit of the particle is an arc of a circle passing through the force center, prove that $n = 5$.

Solution Let D be the diameter of the circle. Consider any position of the particle P. By an elementary property of a circle, we have $\angle OPA = \frac{\pi}{2}$ and $\angle AOX = \pi/4$. Hence

$$r = D\cos\left(\frac{\pi}{4} - \theta\right) \qquad (*)$$

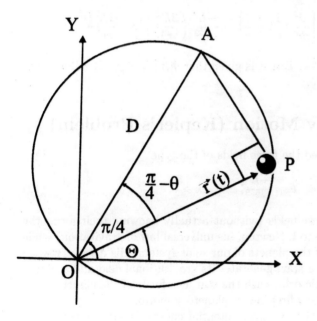

Figure 4.11

Now

$$\frac{d}{d\theta}(\frac{1}{r}) = \frac{d}{d\theta}\left(\frac{\sec\left(\frac{\pi}{4}-\theta\right)}{D}\right) = -\frac{\sin\left(\frac{\pi}{4}-\theta\right)}{D\cos^2\left(\frac{\pi}{4}-\theta\right)}.$$

Consequently,

$$\frac{d^2}{d\theta^2}(\frac{1}{r}) = \frac{d}{d\theta}\left(\frac{-\sin\left(\frac{\pi}{4}-\theta\right)}{D\cos^2\left(\frac{\pi}{4}-\theta\right)}\right) = \frac{\cos^3\left(\frac{\pi}{4}-\theta\right)+2\cos\left(\frac{\pi}{4}-\theta\right)\sin^2\left(\frac{\pi}{4}-\theta\right)}{D\cos^4\left(\frac{\pi}{4}-\theta\right)}$$

$$= \frac{\cos^2\left(\frac{\pi}{4}-\theta\right)+\sin^2\left(\frac{\pi}{4}-\theta\right)}{D\cos^3\left(\frac{\pi}{4}-\theta\right)}$$

Therefore,

$$\begin{aligned}
\frac{d^2}{d\theta^2}\left(\frac{1}{r}\right)+\frac{1}{r} &= \frac{\cos^2\left(\frac{\pi}{4}-\theta\right)+2\sin^2\left(\frac{\pi}{4}-\theta\right)}{D\cos^3\left(\frac{\pi}{4}-\theta\right)}+\frac{1}{D\cos\left(\frac{\pi}{4}-\theta\right)}\\[2mm]
&= \frac{\cos^2\left(\frac{\pi}{4}-\theta\right)+2\sin^2\left(\frac{\pi}{4}-\theta\right)}{D\cos^3\left(\frac{\pi}{4}-\theta\right)}+\frac{1}{D\cos\left(\frac{\pi}{4}-\theta\right)}\\[2mm]
&= \frac{2}{D\cos^3\left(\frac{\pi}{4}\right)}\\[2mm]
&= \frac{2D^2}{r^3} \qquad \text{by }(*).
\end{aligned}$$

Finally, we have

$$\Phi(r) = \frac{-L^2}{mr^2}\left[\frac{d^2}{d\theta^2}(\frac{1}{r})+\frac{1}{r}\right] = \frac{-L^2}{mr^2}\left(\frac{2D^2}{r^3}\right) = \frac{-2L^2D^2}{r^5}$$

Thus, $\Phi(r) = \frac{-2L^2D^2}{r^5}$. But it is given that $\Phi(r) = -\frac{k}{r^n}$. It follows that $n = 5$ (and $k = 2L^2D^2$). $\qquad\square$

4.8   Planetary Motion (Kepler's Problem)

Johannes Kepler studied the force fields of the type

$$F = \frac{-k}{r^2}\vec{e_r}, \qquad k > 0. \tag{12}$$

He came across these fields (without actually knowing their form; the actual form (12) is due to I. Newton, his universal law of gravitation) while studying the motion of the planets of our solar system. His discovery was : A heavenly body (e.g. a star) generates its gravitational field and a planet of it describes an elliptic orbit with the star at a focus of the elliptic orbit. This is known as Kepler's first law of planetory motion.

For the force field (12) we take potential energy function $U(r) = \frac{-k}{r}$. If the particle (i.e. a planet) is moving with angular momentum having magnitude L, then the resulting effective potential energy function W of the particle is $W(r) = \frac{-k}{r} + \frac{L^2}{2mr^2}$.

The graph of W is as shown in Fig. 4.12.

Let E be the total energy of the particle. From the graph (i.e. Fig. 4.11) it is clear that the motion is unbounded if $E \geq 0$; for $E = W(r)$ has only one root which is the minimum value r_{min} of r. Consequently $r_{min} \leq r(t) < \infty$. This explains the unbounded nature of motion of the particle in the case $E \geq 0$.

Note that the function W has a single minimum at $r = \frac{L^2}{mk}$ and the minimum value of W is $W_{\min} = W\left(\frac{L^2}{mk}\right) = \frac{-mk^2}{2L^2}$. Therefore, $E = W_{\min}$ corresponds to a stable circular orbit of radius $= \frac{L^2}{mk}$. Also note that when the total energy E of the particle satisfies $W_{\min} < E < 0$, the equation $E = W(r)$ has precisely two roots.

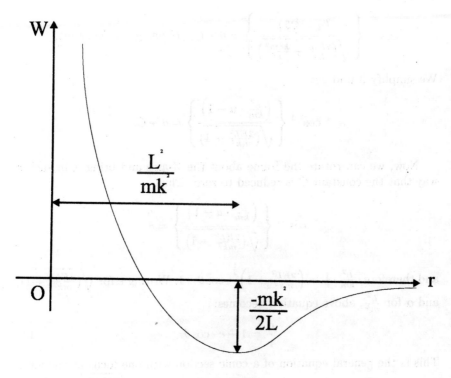

Figure 4.12

In the general situation, we have $E = \frac{m\dot{r}^2}{2} + \frac{L^2}{2mr^2} - \frac{k}{r}$ and therefore, $\dot{r} = \sqrt{\frac{2}{m}\left[E + \frac{k}{r} - \frac{L^2}{2mr^2}\right]}$. Also, $\frac{d\theta}{dt} = \frac{dr}{d\theta} \cdot \frac{d\theta}{dt} = \frac{L}{mr^2} \cdot \frac{dr}{d\theta}$ and consequently

$\frac{L}{mr^2} \cdot \frac{dr}{d\theta} = \sqrt{\frac{2}{m}\left[E + \frac{k}{r} - \frac{L^2}{2mr^2}\right]}$. Equivalently written

$$\frac{\frac{L}{mr^2}\,dr}{\sqrt{\frac{2}{m}\left[E + \frac{k}{r} - \frac{L^2}{2mr^2}\right]}} = d\theta.$$

Putting $u = \frac{1}{r}$, we get

$$\frac{\frac{-L}{m}\,du}{\sqrt{\frac{2}{m}\left[E + ku - \frac{L^2}{2m}u^2\right]}} = d\theta.$$

Rewriting it in the form

$$\frac{-du}{\sqrt{\frac{2Em}{L^2} + \frac{k^2m^2}{L^4} - \left(u - \frac{km}{L^2}\right)^2}} = d\theta.$$

We integrate it to get

$$\cos^{-1}\left\{\frac{\left(u - \frac{km}{L^2}\right)}{\sqrt{\left(\frac{2Em}{L^2} + \frac{k^2m^2}{L^4}\right)}}\right\} = \theta + C, \quad C \text{ being   a constant,}$$

We simplify it and get

$$\cos^{-1}\left\{\frac{\left(\frac{L^2}{km} \cdot u - 1\right)}{\sqrt{\left(\frac{2EL^2}{mk^2} + 1\right)}}\right\} = \theta + C.$$

Now, we can rotate the frame about the Z-axis and bring it in such a way that the constant C is reduced to zero. Thus,

$$\cos^{-1}\left\{\frac{\left(\frac{L^2}{km} \cdot u - 1\right)}{\sqrt{\left(\frac{2EL^2}{mk^2} + 1\right)}}\right\} = \theta$$

and therefore, $\frac{L^2}{km} \cdot \frac{1}{r} = \left(\frac{2EL^2}{mk^2} + 1\right)^{\frac{1}{2}} \cdot \cos\theta + 1$. Writing e for $\sqrt{\left(\frac{2EL^2}{mk^2} + 1\right)}$, and α for $\frac{L^2}{mk}$, above equation becomes:

$$\frac{\alpha}{r} = 1 + e \cdot \cos\theta. \tag{13}$$

This is the general equation of a conic section with one focus at the force center 0. The conic section has eccentricity $e = \sqrt{\left(\frac{2EL^2}{mk^2} + 1\right)}$ and $\alpha = \frac{L^2}{mk}$ as its semi-latus rectum. Note how e and α depend on the dynamic parameters E and L. Consequently, for a given valued of L, the shape of the orbit is determined by the total energy E of the particle. In fact, (according to elementary coordinate geometry) we have the following classification:

(I) If $E > 0$, then $e > 1$ and the orbit is one branch of hyperbola.

(II) If $E = 0$, then $e = 1$ and the orbit is a parabola.

(III) If $0 > E > \frac{-mk^2}{2L^2}$, then $0 < e < 1$ and the orbit is an ellipse.

(IV) If $E = W_{\min} = \frac{-mk^2}{2L^2}$ then $e = 0$ and the orbit is stable circular.

The planetory motion corresponds to the energy E of the planet in the in the range $\frac{-mk^2}{2L^2} \le E < 0$ which, according to the above classification is an elliptic or a circle. This is Kepler's first law: Planets move in elliptic orbits.

Let us consider an elliptic orbit of a planet. Let $2a$ and $2b$ be the lengths of its major axis and minor axis respectively. Then a and b are given by

$$\left.\begin{array}{l} a = \frac{\alpha}{1-e^2} = \frac{k}{-2E} = \frac{k}{2|E|}, \text{this is so because} E < 0 \\ b = \frac{\alpha}{\sqrt{1-e^2}} = \frac{L}{\sqrt{2m|E|}} \end{array}\right\} \quad (14)$$

Also, the area of the ellipse is $A = \pi a \cdot b$. Using the relations (14) we get

$$A = \frac{\pi k}{2\,|\,E\,|} \cdot \frac{L}{\sqrt{2m\,|\,E\,|}} = \frac{\pi k L}{\sqrt{(8m)}\,|\,E\,|^{3/2}} \quad (15)$$

Since an elliptic orbit is a closed curve, the motion of the particle is periodic. Let τ be the period of motion. Thus, τ is the time required by the particle (or the planet) to complete one round of its elliptic orbit.

Recall, we have proved : $\frac{d}{dt}A(t) = \frac{L}{2m}$ and hence $dA = \frac{L}{2m}dt$. We want to integrating the last infinitesimal equation. Note that integration over the period τ corresponds to calculating the area of the ellipse. Therefore $A = \int dA = \frac{L}{2m}\int_0^\tau dt = \frac{L}{2m}\tau$. Combining this result with (15), we get $\frac{\pi k L}{\sqrt{(8m)}|E|^{3/2}} = \frac{L}{2m}\tau$. This gives $\tau = \sqrt{\frac{m}{2}} \cdot \frac{\pi k}{|E|^{3/2}}$ or equivalently,

$$\tau^2 = \left(\frac{m}{2}\right)\frac{\pi^2 k^2}{|\,E\,|^3}. \quad (16)$$

Also, using (14), the above can be rewritten as $\tau^2 = \frac{4\pi^2 m}{k}a^3$. This proves the propertionality relation; $\tau^2 \alpha a^3$ which is Kepler's *third law of planetory motion*: Planets move along their elliptic orbits in such a way that the cube of length of the semi-major axis is proportional to the square of the period of the motion:

Example 7 A particle moves in an elliptic orbit under an inverse square law force. If the ratio of the maximum angular velocity to the minimum angular velocity of the particle is η, show that the eccentricity e of the orbit is given by $e = \frac{\sqrt{\eta}-1}{\sqrt{\eta}+1}$.

Solution

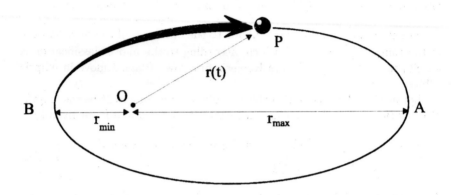

Figure 4.13

Let L be the magnitude of the angular momentum. Now, we have $\dot{\theta} = \frac{L}{mr^2}$. Consequently $\dot{\theta}_{max} = \frac{1}{m(r_{min})^2}$ and $\dot{\theta}_{min} = \frac{L}{m(r_{max})^2}$.

Now, $r_{max} = \ell(OA) = a(1 + r)$ and $r_{min} = \ell(OB) = a(1 - e)$. Therefore, $\dot{\theta}_{max} = \frac{1}{ma^2(1-e)^2}$ and $\dot{\theta}_{min} = \frac{1}{ma^2(1+e)^2}$. Consequently, $\eta = \frac{\dot{\theta}_{max}}{\dot{\theta}_{min}} = \frac{1}{ma^2(1-e)^2} \cdot \frac{ma^2(1+e)^2}{1}$ which gives $\eta = \frac{(1+e)^2}{(1-e)^2}$ or $\sqrt{\eta} = \frac{1+e}{1-e}$ which gives $e = \frac{\sqrt{\eta}-1}{\sqrt{\eta}+1}$. $\qquad\qquad\square$

Example 8 The particle P is moving along an elliptical orbit in an attractive inverse square law force field. Prove that the product of the maximum and minimum speeds of its is $\frac{2}{m} \mid E \mid$.

Solution Let the force field be given by $\phi(r)\vec{e}_r = \frac{-k}{r^2}\vec{e}_r$, $\quad k > 0$. Then the potential energy function U is given by $v(\vec{r}) = -\frac{k}{r}$ and so, the total energy E of the particle is $E = \frac{m}{2}\dot{r}^2 - \frac{k}{r}$. This gives $v = \dot{r} = \sqrt{\frac{2}{m}[E + \frac{k}{r}]}$. Consequently, $U_{max} = \sqrt{\frac{2}{m}[E + \frac{k}{r_{min}}]}$ and $v_{min} = \sqrt{\frac{2}{m}[E + \frac{k}{r_{min}}]}$. Recall, $r_{max} = a(1 + e)$, $r_{min} = a(1 - e)$ and $a = \frac{\alpha}{1-e^2}$. Therefore, $r_{max} = \frac{\alpha}{1-e}$ and $r_{min} = \frac{\alpha}{1+e}$ and consequently,

$$v_{max} \cdot v_{min} = \frac{2}{m}\sqrt{\left(E + k\frac{(1-e)}{\alpha}\right)} \cdot \sqrt{\left(E + \frac{k(1+e)}{\alpha}\right)}.$$

$$= \sqrt{\left(E + \frac{k}{\alpha} - \frac{ke}{\alpha}\right)\left(E + \frac{k}{\alpha} - \frac{ke}{\alpha}\right)}$$

$$= \frac{2}{m}\sqrt{\left(E^2 + \frac{2k}{\alpha}E + \frac{k^2}{\alpha^2} - \frac{k^2e^2}{\alpha^2}\right)}$$

$$= \frac{2}{m}\sqrt{\left(D^2 + \frac{2kE}{\alpha} + \frac{k^2}{\alpha^2}(1-e^2)\right)}$$

$$= \frac{2}{m}\sqrt{\left(E^2 + \frac{2kE}{\alpha} + \frac{k^2}{\alpha^2} \cdot \frac{\alpha}{a}\right)}.$$

$$= \frac{2}{m}\sqrt{\left(E^2 + \frac{2kE}{\alpha} + \frac{k^2}{\alpha a}\right)}.$$

$$= \frac{2}{m}\sqrt{\left(E^2 + 2k \cdot E \cdot \frac{km}{L^2} - \frac{2k \cdot E}{L^2}\right)}$$

$$= \frac{2}{m}\,|E|.$$

The last step being obtained by substituting $\frac{1}{\alpha} = \frac{km}{L^2}$ and $a = \frac{-k}{2E}$. Therefore, $v_{max} \cdot v_{min} = \frac{2}{m}\,|E|$. □.

Example 9 A particle moves in a circular orbit under the influence of the force field $\frac{-k}{r^2}\vec{e_r}$. If suddenly k decreases to half of its original value, show that the orbit of the particle becomes a parabola.

Solution Let a be the radius of the circular orbit and let E be the total energy of the particle before the change in the force law takes place. Along the circular orbit, we have $r = a$, $\dot{\theta}$ = constant and therefore, $v = a\dot{\theta}$ the speed of the particle, is also constant. Now, since the particle is moving in the circular orbit, the centrifugal force balances the (attractive) central force. As a consequence, the total energy of the particle is its kinetic energy.

Now centrifugal force = attractive force gives $\frac{mv^2}{a} = \frac{k}{a^2}$ and hence $v^2 = \frac{k}{ma}$. Let E' be the total energy of the particle *after* the change in the force law occurs. Let A be the point on the circular orbit such that the change in force law occurs when the particle is at A. Then the total energy E' at the moment is

$$E' = \frac{m}{2}v^2 - \frac{k}{2a} = \frac{mk}{2ma} - \frac{k}{2a} = 0.$$

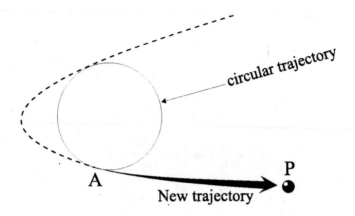

Figure 4.14

Thus, after the change in the force law takes place, the partial moves with the total energy zero in the new central force field $\frac{-k}{2r^2}\vec{e_r}$. The condition $E' = 0$ characterizes a parabola in such a (inverse square force) field and hence, the new trajectory must be a parabola. $\square$

Example 10 Suppose, a particle P of mass m is moving along a circular orbit of radius a under the inverse square type of force. Suppose, the speed of the particle is increased suddenly by a fraction $\lambda(\lambda < \sqrt{2} - 1)$ without changing its direction of motion: Prove that the subsequent motion is elliptic with semi minor axis $a' = a(1 - 2\lambda - \lambda^2)^{-1}$.

Solution Let the force be given by $\Phi(r)\vec{e_r} = \frac{-k}{r^2}\vec{e_r}$. Along the circular orbit, we have $\frac{mv^2}{a} = \frac{k}{a^2}$ which gives $v^2 = \frac{k}{ma}$. Now the speed is increased to the value $(1 + \lambda)v$. Therefore, the new energy E' is given by

$$
\begin{aligned}
E' &= \frac{m}{2}(1+\lambda)^2 v^2 - \frac{k}{a} = \frac{m}{2}(1+\lambda)^2 \frac{k}{ma} - \frac{k}{a}, \text{ using } v^2 = \frac{k}{ma} \\
&= \frac{k}{2a}\left((1+\lambda)^2 - 2\right) = \frac{k}{2a}(1 + 2\lambda + \lambda^2 - 2) = \frac{k}{2a}(\lambda^2 + 2\lambda - 1)
\end{aligned}
$$

Since $\lambda < \sqrt{2-1}$ (given) we get that $E' < 0$. Therefore, the new orbit is an ellipse, its semi-major axis a' being given by

$$
a' = \frac{k}{2\,|\,E'\,|} = \frac{k}{-2E'} = \frac{2ak}{-2k(\lambda^2 + 2\lambda - 1)} = \frac{a}{(1 - 2\lambda - \lambda^2)}. \qquad \square
$$

Example 11: For the central force field given below

$$
F(\vec{r}) = -\left(\frac{k}{r^2} + \frac{k'}{r^4}\right)\vec{e_r};
$$

show that, if $\rho > 0$, is a constant satisfying $\rho^2 k > k'$, then a particle can move in a stable circular orbit of radius ρ.

Solution: Let m be the mass of the particle, and let L be the magnitude of its angular momentum. Then its effective potential energy function W is given by

$$W(r) = \frac{L^2}{2mr^2} + U(r)$$

where $U(r) = - \left(\frac{k}{r} + \frac{k'}{3r^3} \right)$ is the potential energy of the given force field. Now the following two conditions are sufficient for a stable circular orbit of radius ρ: (i) $\frac{dW}{dr}(\rho) = 0$ and (ii) $\frac{d^2W}{dr^2}(\rho) > 0$. [Recall, conditions (i) and (ii) express the condition that the effective potential energy function W has a local minimum at $r = \rho$].

Now, $\frac{dW}{dr}(\rho) = -\frac{L^2}{m\rho^3} + \frac{k}{\rho^2} + \frac{k'}{\rho^4}$. Therefore, $\frac{dW}{dr}(\rho) = 0$ implies that $L^2 = mk\rho + \frac{mk'}{\rho}$. Also,

$$\begin{aligned}
\frac{d^2W}{dr^2}(\rho) &= \frac{3L^2}{m\rho^4} - \frac{2k}{\rho^3} - \frac{4k'}{\rho^5} \\
&= \frac{1}{\rho^3} \left[\frac{3}{\rho m}(mk\rho + \frac{mk'}{\rho}) - 2k - \frac{4k'}{\rho^2} \right] \\
&= \frac{1}{\rho^3} \left[k - \frac{k'}{\rho^2} \right]
\end{aligned}$$

Hence $\frac{d^2W}{dr^2}(\rho) > 0$ if and only if $k - \frac{k'}{\rho^2} > 0$.

Thus, (given) $\rho^2 k > k'$ implies that the effective potential energy W has a local minimum at $r = \rho$ and hence the particle can perform stable circular motion.   □

Example 12 A particle of mass m describes an ellipse under an attractive force $\frac{-mk}{r^2}\vec{e_r}$. Suppose, the particle is set in motion with speed v_o when it is at a distance ρ from the force center. Prove that the period τ of motion is $\frac{2\pi}{\sqrt{k}} \left[\frac{2}{\rho} - \frac{v_o^2}{k} \right]^{\frac{-3}{2}}$

Solution Note that the potential energy $U(r)$ of the force field is $\frac{-mk}{r}$. Therefore, the total energy is $E = \frac{mv^2}{2} - \frac{mk}{r}$. Calculating it at the instant described in the example, we get.

$$E = \frac{m}{2}v_0^2 - \frac{mk}{\rho} \qquad (*)$$

[Note that in this problem, we are having everywhere, the constant mk in place of the constant k described in the Keplerian problem.] Now according to the derivation (16) the period τ is given by

$$\tau = \sqrt{(\frac{m}{2})} \cdot \frac{\pi mk}{|E|^{\frac{3}{2}}} \qquad (**)$$

Noting that E is negative (since the orbit is elliptic) we have

$$| E |= \frac{mk}{\rho} - \frac{m}{2}v_o^2 \qquad\qquad (* * *)$$

Substituting $(* * *)$ in $(**)$ and simplifying the resulting expression, we get the desired expression for τ. $\qquad\qquad\qquad\qquad\qquad\qquad\qquad$ □

EXERCISES

1. Find the force field which allows a particle to move in a spiral orbit given by $r = k\theta^2$, k being a constant.

2. A particle moves in the orbit $r = a \cos 2\theta$, under an attractive central force field. Show that the magnitude of the force field is proportional to $\frac{8a^2 - 3r^2}{r^5}$.

3. A particle is moving in a central force field of magnitude $-\frac{k}{r^3}$. Show that it is possible to choose the values of the total energy E and the angular momentum L in such a way that the orbit of the particle is of the form $r = ae^{b\theta}$.

4. A particle P of unit mass is acted upon by an attractive central force field of magnitude $w^2 r + \frac{w^2 a^3}{r^2}$. If P is projected from a point A at a distance a from the force center with velocity $\frac{4aw}{\sqrt{3}}$, perpendicular to the position vector, show that subsequently, the distance of P from the force center varies between $r = a$ and $r = 2a$ and find the speed when $r = 2a$.

5. Two particles A and B, each of mass m are connected by an inextensible string of length $2a$, which passes through a small, smooth hole in a smooth horizontal table. The particle A is free to slide smoothly on the table and B hangs from the hole. Initially OB is of length a and the particle A is projected from rest with a speed $(\frac{8ga}{3})^{\frac{1}{2}}$ in a direction perpendicular to OA. Show that in the subsequent motion the particle B will start climbing the hole and will just reach it.

6. A particle P having mass m is describing an ellipse of having major and minor axes $2a$ and $2b$ respectively about the center of the force which is located at the center of the ellipse. When the particle reaches the end of the major axis, it strikes and coal eases with a particle Q of mass mn which is at rest. The central attraction per unit mass is

unchanged. Prove that the new orbit is an ellipse, having new major axis $2a$ and the new minor axis $\frac{2b}{(n+1)}$.

7. A particle of mass m describes an ellipse under a force of magnitude $\frac{m\mu}{r^2}$ directed towards the focus. Show that if the particle is moving with speed v_o when it is at a distance d from the center of force, the period of motion is

$$\frac{2\pi}{\sqrt{\mu}} \left[\frac{2}{d} - \frac{v_o^2}{\mu} \right]^{\frac{-3}{2}} .$$

8. A particle of mass mk is describing a parabolic orbit of latus rectum $4a$ under the inverse square law $\frac{-k}{r^2}\vec{e_r}$. At one end of the latus rectum, it meets and coaleases with another particle of mass $m \cdot n$ at rest. Show that the composite particle will trace out an elliptic orbit of eccentricity e given by the equation

$$(n+1)^2(1-e^2) = 2n(n+2).$$

Chapter 5

System of Particles

" Nature, it seems, is the popular name for millards and millards and millards of particles playing their infinite game of billiards and billiards and billiards".

Piethein

5.1 Introduction

In this chapter we study the motion of a mechanical system consisting of finitely many particles. Though the number of particles in the system is assumed to be finite, it can be arbitrarily large. We will see that all the notions and results discussed earlier for a single particle can be extended to such a system.

An important role is played in the analysis of motion of the system by the familiar point, namely its *center of mass*. In general, it is a point moving with the system. In fact, it will be shown that it moves exactly as if it were a material particle. Consequently our familiarity with the motion of a single particle (gained in the previous chapters) enables us to study the motion of the center of mass in terms of the associated fictitious material particle. This is our first step in studying the motion of the system of particles.

The second step is then to study the motion of the constituent particles *relative to* the center of mass.

This method of splitting the motion in two parts proves beneficial essentially because of the fact that many dynamical quantities associated with the particle system are expressible as the sum of two terms, the first being related to the motion of the center of mass and the second pertaining to the motion of the particles relative to the center of mass.

As a particular case, we study the *two body problem*. This is a problem about the motion of a pair of particles, the only forces acting on them being forces of their mutual interactions. It will be proved that we can associate a third particle (not the center of mass) and study the motion of the two particle in terms of the motion of the single hypothetical particle.

5.2   A System of Particles

Let
$$\{P_1, P_2, \cdots P_n\} \tag{1}$$

be a finite family of particles having their respective masses $m_1, m_2, \cdots, m_n$. In the following we will refer to this family as the *system*.

As usual, let $\mathcal{F}$ be a stationary frame of reference. We use it to identify the physical space V with $\mathbb{R}^3$. All the vector quantities associated with the system will be considered to be taken in reference to the frame $\mathcal{F}$.

Let Ω_ℓ be the region in V in which the particle P_ℓ of the system is moving. Suppose, the motion of the system is observed during the time interval I. Then, as in the earlier single particle case, the motion of each particle P_ℓ is a (time parameterized) curve

$$c_\ell : I \longrightarrow \Omega_\ell \tag{2}$$

We denote the position vector of $c_\ell(t)$ by $\vec{r}_\ell(t)$ and its resolution with respect to $\mathcal{F}$ by

$$\vec{r}_\ell(t) = x_\ell(t)\vec{i} + y_\ell(t)\vec{j} + z_\ell(t)\vec{k} \tag{3}$$

Often, to make our notations less cumbersome, we drop the time parameter t and write simply $\vec{r}_\ell$ for $\vec{r}_\ell(t), x_\ell$ for $x_\ell(t)$ etc. We want to use another notational simplification: Instead of writing $\sum_{\ell=1}^{n} m_\ell$, we will write $\sum_\ell m_\ell$, instead of $\sum_{\ell=1}^{n} \vec{p}_\ell$ only $\sum_\ell \vec{p}_\ell$ and so on. (This convention is applicable whenever a summation extends over all the particles of the system.)

If the reader is familiar with the notion of Cartesian product of sets and maps then he may combine instantaneous position vectors. $\vec{r}_1(t)$ of P_1 in Ω_1, $\vec{r}_2(t)$ of P_2 in $\Omega_2 \cdots \vec{r}_n(t)$ of P_n in Ω_n in a single point $(\vec{r}_1(t), \vec{r}_2(t), \cdots, \vec{r}_n(t))$ of the Cartesian product $\Omega_1 \times \Omega_2 \times \cdots \Omega_n$. Similarly, the set $\{c_\ell : I \longrightarrow \Omega_\ell\}_{\ell=1}^{n}$ of trajectories of the particles may be combined in a single curve:

$$c : I \longrightarrow \Omega_1 \times \Omega_2 \times \cdots \times \Omega_n; t \longmapsto (\vec{r}_1(t), \vec{r}_2(t), \cdots, \vec{r}_n(t)). \tag{4}$$

In the following, by *trajectory of the system*, we will mean the curve (4).

We consider the forces acting on the particles of the system now. The net force acting on the particle P_ℓ is the sum of the following two types of forces:

(a) The forces of interaction between the particles of the system: Let $\vec{F}_{\ell j}(\vec{r}_\ell, \vec{r}_j)$ be the force on P_ℓ exerted by P_j. The sum $\sum\limits_{j(j \neq \ell)} \vec{F}_{\ell j}(\vec{r}_\ell, \vec{r}_j)$ is the total internal force acting on P_ℓ (We do not allow the summoned $\vec{F}_{\ell\ell}(\vec{r}_\ell, \vec{r}_\ell)$ because it is zero, the particle P_ℓ does not exert a force on itself). Recall, according to Newton's third law of action and reaction, we have $F_{\ell j} \equiv -F_{j\ell}$. (and consequently, $F_{\ell\ell} = 0$).

(b) The external force $\vec{F}_\ell^{(e)}(\vec{r}_\ell)$ arising from the interaction of P_ℓ with the rest of the matter in the universe.

Therefore, the total force acting on P_ℓ is

$$F_\ell = F_\ell^{(e)} + \sum_{\ell \neq j} F_{\ell j} \tag{5}$$

Clearly $F_\ell = F_\ell(\vec{r}_1, \vec{r}_2, \cdots \vec{r}_n)$.

Now the equation of motion of P_ℓ is

$$m_\ell \ddot{\vec{r}}_\ell = \vec{F}_\ell(\vec{r}_1, \vec{r}_2, \cdots, \vec{r}_n). \tag{6}$$

5.3 The Center of Mass

We consider the point which is given in terms of its position vector denoted by $\vec{R} = \vec{R}(t)$ where

$$\vec{R}(t) = \frac{\left(\sum_\ell m_\ell \vec{r}_\ell(t)\right)}{\left(\sum_\ell m_\ell\right)} \tag{7}$$

Definition 1 The point represented by $\vec{R}(t)$ is the **center of mass** of the system. We will denote the center of mass of the system by C.

We will write M for $\sum\limits_\ell m_\ell$ and refer to it as the *total mass* of the system (1). Now equation (7) gives

$$M\vec{R}(t) = \sum_\ell m_\ell \vec{r}_\ell(t). \tag{8}$$

Differentiating (8) with respect to time, we get

$$M\dot{\vec{R}}(t) = \sum_\ell m_\ell \dot{\vec{r}}_\ell(t) \tag{9}$$

Clearly, $\dot{\vec{R}}(t)$ is the instantaneous velocity of the center of mass. We denote it by $\vec{V}(t)$.

Differentiating (9), we get

$$M\ddot{\vec{R}}(t) = \sum_{\ell} m_{\ell}\ddot{\vec{r}}_{\ell}(t) = \sum_{\ell} F_{\ell}.$$

Thus we get the equation

$$M\ddot{\vec{R}} = \sum_{\ell} F_{\ell}. \tag{10}$$

Left hand sides of equations (9) and (10) suggest that we should consider a hypothetical particle- let us call it P-having mass M and being located at the center of mass all the time (so that the motion of the particle P is the same as the motion of the center of mass C.) Then $M\dot{\vec{R}}(t)$ is the linear momentum of P, $\ddot{\vec{R}}(t)$ is its acceleration and $\sum_{\ell} F_{\ell}$ can be considered as the total force acting on it. Now equation (10) has the interpretation of being the equation of motion of the particle P which has mass M and which is acted upon by the force $\sum_{\ell} F_{\ell}$.

Moreover, the equations $F_{\ell j} = -F_{j\ell}$ between the interacting forces make them disappear from equation (10). This is seen below:

$$\begin{aligned}
\sum_{\ell} F_{\ell} &= \sum_{\ell}\left\{ F_{\ell}^{(e)} + \sum_{j(\neq \ell)} F_{\ell j} \right\} \\
&= \sum_{\ell} F_{\ell}^{(e)} + \sum_{j\neq \ell} F_{\ell j} \\
&= \sum_{\ell} F_{\ell}^{(e)} + \sum_{\ell>j} F_{\ell j} + \sum_{\ell<j} F_{\ell j} \\
&= \sum_{\ell} F_{\ell}^{(e)} + \sum_{\ell>j} F_{\ell j} + \sum_{j<\ell} F_{j\ell}
\end{aligned}$$

where in the last summation we have interchanged the (dummy) summation indices j and ℓ. Next, replacing $F_{j\ell}$ by $-F_{\ell j}$ there, we get

$$\sum_{\ell} F_{\ell} = \sum_{\ell} F_{\ell}^{(e)} + \sum_{\ell>j} F_{\ell j} + \sum_{\ell>j} (-F_{\ell j}) = \sum_{\ell} F_{\ell}^{(e)} + 0.$$

Thus, $\sum_{\ell} F_{\ell} = \sum_{\ell} F_{\ell}^{(e)}$ and consequently equation (10) becomes

$$M\ddot{\vec{R}}(t) = \sum_{\ell} F_{\ell}^{(e)} \tag{11}$$

We have proved the following proposition.

Proposition 1 The center of mass C of the system moves as if it were a particle having mass equal to the total mass of the system (1), acted upon by a force equal to the vector sum of the external forces acting on the particles of the system.

In a situation where the sum $\sum_{\ell} F_{\ell}^{(e)}$ is zero equation (11) reduces to

$M\ddot{\vec{R}}(t) = 0$ giving constancy of the velocity $\vec{V}(t) = \dot{\vec{R}}(t)$ of the center of mass. Hence we have the following corollary to **Proposition 1:**

Corollary If the vector sum of all the forces acting on the system vanishes identically then the center of mass of the system either remains stationary or moves in a straight line with constant speed.

The above discussion makes it clear that the forces of interaction $\{F_{\ell j}\}$ have no influence on the motion of the center of mass.

Sometimes, the system (1) is said to be *closed* if all the external forces are zero: $F_{\ell}^{(e)} \equiv 0; 1 \leq \ell \leq n$.

Thus the above corollary implies that the center of mass of a closed mechanical system moves in a straight line with constant speed.

5.4 Some More Definitions

We define a number of dynamical quantities for the particle system (1) linear momentum, angular momentum, kinetic energy, potential energy, total energy, and so on.

Let us recall the notations we have been using and also introduce some more:

(i) $\vec{r_{\ell}}(t)$ is the instantaneous position vector of P_{ℓ}, $\vec{v_{\ell}}(t) = \dot{\vec{r_{\ell}}}(t)$ its velocity $\vec{a_{\ell}}(t) = \ddot{\vec{r_{\ell}}}(t)$, the acceleration (all taken with respect to the stationary frame $\mathcal{F}$.)

(ii) $\vec{R}(t)$ is the instantaneous position vector of the center of mass, $\dot{\vec{R}}(t) = \vec{V}(t)$, its velocity and $\ddot{\vec{R}}(t)$, the acceleration.

As we intend to study the motion of the particles relative to the center of mass, we introduce the following notations:

(iii) $\vec{s_{\ell}}(t)$ is the instantaneous position vector *relative to* C, $\dot{\vec{s_{\ell}}}(t)$ and $\ddot{\vec{s_{\ell}}}(t)$ are the velocity and acceleration of it relative to C.

Obviously, we have

$$\vec{r_{\ell}}(t) = \vec{R}(t) + \vec{s_{\ell}}(t) \tag{12}$$

Now, we have

$$\begin{aligned} M\vec{R} &= \sum_{\ell} m_{\ell}\vec{r_{\ell}} = \sum_{\ell} m_{\ell}(\vec{R} + \vec{s_{\ell}}) \\ &= \sum_{\ell} m_{\ell}\vec{R} + \sum_{\ell} m_{\ell}\vec{s_{\ell}} = M\vec{R} + \sum_{\ell} m_{\ell}\vec{s_{\ell}} \end{aligned}$$

Thus we get $M\vec{R} = M\vec{R} + \sum_\ell m_\ell \vec{s}_\ell$ and hence

$$\sum_\ell m_\ell \vec{s}_\ell = 0 \tag{13}$$

Differentiating equation (13) twice with respect to time, we get

$$(a) \sum_\ell m_\ell \dot{\vec{s}}_\ell = 0 \text{ and } (b) \sum_\ell m_\ell \ddot{\vec{s}}_\ell = 0 \tag{14}$$

Now we define the following terms.

Definition 2 The sum $\sum_\ell m_\ell \dot{\vec{r}}_\ell(t) = \sum_\ell \vec{p}_\ell(t)$ is the instantaneous **linear momentum** of the system.

We denote it by $\vec{p}(t)$. Let A be a stationary point and let $\vec{a}$ be its position vector.

Definition 3 The sum $\sum_\ell m_\ell (\vec{r}_\ell(t) - \vec{a}) \times \vec{v}_\ell$ is the instantaneous **angular momentum** of the system about the point A.

We denote it by $\vec{L}_A(t)$.(Often when the point A is clear from the context, we drop the suffix A to make the notation less cumbersome.)

Definition 4 The scalar quantity $T = T(t) = \frac{1}{2} \sum_\ell m_\ell \|\vec{v}_\ell(t)\|^2$ is the instantaneous **kinetic energy** of the system.

Sometimes the kinetic energy is treated as the functions of the velocities of the constituent particles of the system

$$T = T(\vec{v}_1, \vec{v}_2, \cdots, \vec{v}_n) = \frac{1}{2} \sum_\ell m_\ell \|\vec{v}_\ell\|^2.$$

Definition 5 The particle system is said to be **conservative** or equivalently, the forces acting on it are said to be conservative if there exists a function

$$U : \Omega_1 \times \Omega_2 \times \cdots \times \Omega_n \longrightarrow \mathbb{R}; (r_1, r_2, \cdots r_n) \longrightarrow U(\vec{r}_1, \vec{r}_2, \cdots, \vec{r}_n)$$

which satisfies the differential equations

$$F_\ell = -\left[\frac{\partial U}{\partial x^\ell} \vec{i} + \frac{\partial U}{\partial y^\ell} \vec{j} + \frac{\partial U}{\partial z^\ell} \vec{k} \right] ; 1 \leq \ell \leq n.$$

We will use the notation $\text{grad}_\ell(U)$ for the gradient $\frac{\partial U}{\partial x^\ell}\vec{i} + \frac{\partial U}{\partial y^\ell}\vec{j} + \frac{\partial U}{\partial z^\ell}\vec{k}$. Thus $F_\ell = -\text{grad}_{(\ell)}(U)$.

Definition 6 A function $U : \Omega_1 \times \Omega_2 \times \cdots \times \Omega_n \longrightarrow \mathbb{R}$ satisfying $F_\ell = -$ grad $_{(\ell)}(U)$ is called the **potential energy** function of the particle system.

Recall, in **Chapter 3** we considered the work $W(c_\ell)$ done by the force F_ℓ in taking the particle P_ℓ from $c_\ell(a)$ to $c_\ell(b)$ along the path $c_\ell : I = [a, b] \longrightarrow \Omega_\ell$.

Definition 7 The sum $\sum_\ell W(c_\ell)$ is the **work** done by the forces on the system.

We denote it by $W(c)$.

Again, as in the case of a single particle, the work $W(c)$ may depend only on the end-points of the trajectory c of the system. If this happens for every trajectory of the system, then it can be shown that the system is conservative. Moreover, in that case, we can define the potential energy function in terms of the work leading to the usual equation

$$U(c(b)) - U(c(a)) = -W(c). \tag{15}$$

Because we will not use these ideas very much we have not derived any results related to the concept of work. The reader may work out all the details by himself and obtain the equation (15).

We assume from now-onwards that the particle system is conservative. Its potential energy function will be denoted by U.

Definition 8 The sum function $T + U$ is the **total energy** function of the particle system. We denote it by E.

Since the kinetic energy T is a function of the velocities $(\vec{v}_1 \vec{v}_2 \cdots \vec{v}_n)$ of the particles of the system while the potential energy U is a function of their position vectors $(\vec{r}_1, \vec{r}_2 \cdots \vec{r}_n)$, the total energy is a function of both, the positions and velocities:

$$E = E(\vec{r}_1, \vec{r}_2, \cdots, \vec{r}_n \vec{v}_1 \cdots \vec{v}_n).$$

On the other hand, since a particular trajectory of the system is time parametrized, the total energy along the trajectory becomes a function of time:

$$E = E\left(\vec{r}_1(t), \vec{r}_2(t) \cdots \vec{r}_n(t), \vec{v}_1(t), \vec{v}_2(t) \cdots \vec{v}_n(t)\right) = E(t).$$

Again, as in case of a single particle, the above function $t \longmapsto E(t)$ will be shown to be constant along each of the trajectory of the system.

Let A be a stationary point in space with $\vec{a}$ as its position vector.

Definition 9 The **torque** about A of the forces acting on the system (while at $(\vec{r}_1, \vec{r}_2 \cdots, \vec{r}_n)$) is the vector $\sum (\vec{r}_\ell - \vec{a}) \times F_\ell$.

We denote it by $N_A(\vec{r}_1 \cdots \vec{r}_n)$. Thus we get a vector valued function $N_A : \Omega_1 \times \Omega_2 \times \cdots \times \Omega_n \longrightarrow \mathbb{R}^3$.

Having defined these concepts, we prove a number of basic but simple results now.

We have

$$\vec{p} = \sum_\ell m_\ell \vec{v}_\ell = \sum_\ell m_\ell (\dot{\vec{R}} + \dot{\vec{s}}_\ell) = \sum_\ell m_\ell \dot{\vec{s}}_\ell + M\dot{\vec{R}}$$

$M\dot{\vec{R}}$ is the linear momentum of the center of mass (that is, that of the fictitious particle of mass M associated with the center of mass). Also $\sum_{\ell} m_{\ell}\dot{\vec{s}}_{\ell}$ is the linear momentum of the system when we consider its motion *relative to* the center of mass. Thus we have proved the following proposition:

Proposition 2 The linear momentum of the particle system (1) is the sum of the following two terms: (1) Linear momentum of the center of mass and (2) Linear momentum of the system relative to the center of mass.

Next, we consider the angular momentum of the system about the origin of $\mathcal{F}$, and the same taken about the center of mass and obtain a relation between them

$$
\begin{aligned}
\vec{L}_o &= \sum_{\ell} m_{\ell}\vec{r}_{\ell} \times \dot{\vec{r}}_{\ell} = \sum_{\ell} m_{\ell}\left(\vec{R} + \vec{s}_{\ell}\right) \times \left(\dot{\vec{R}} + s_{\ell}\right) \\
&= \left(\sum_{\ell} m_{\ell}\right)\vec{R} \times \dot{\vec{R}} + \vec{R} \times \left(\sum_{\ell} m_{\ell}\dot{\vec{s}}_{\ell}\right) + \left(\sum_{\ell} m_{\ell}\vec{s}_{\ell}\right) \times \dot{\vec{R}} + \sum_{\ell} m_{\ell}\vec{s}_{\ell} \times \dot{\vec{s}}_{\ell} \\
&= M\vec{R} \times \dot{\vec{R}} + 0 + 0 + \sum m_{\ell}\vec{s}_{\ell} \times \dot{\vec{s}}_{\ell} \text{ using (14) and (15a)} \\
&= \vec{L}(C) + \vec{L}_c
\end{aligned}
$$

where $\vec{L}(C)$ denotes the angular momentum of the center of mass (about the origin of $\mathcal{F}$) and the angular momentum of the system relative to the center of mass.

We summarize the above result in the following proposition

Proposition 3 The angular momentum $\vec{L}_o$ of the system about the origin of $\mathcal{F}$ is the sum of the following two terms
(1) The angular momentum $\vec{L}_c$ of the system, about the center of mass.
(2) The angular momentum $\vec{L}(C)$ of the center of mass taken about the origin of $\mathcal{F}$.

Finally, we prove a similar result about the kinetic energy. We have

$$
\begin{aligned}
T &= \frac{1}{2}\sum_{\ell} m_\ell \|\dot{\vec{r}}_\ell\|^2 \\
&= \frac{1}{2}\sum_{\ell} m_\ell (\dot{\vec{R}} + \dot{\vec{s}}_\ell) \cdot (\dot{\vec{R}} + \dot{\vec{s}}) \\
&= \frac{1}{2}\left(\sum_{\ell} m_\ell\right)\dot{\vec{R}} \times \dot{\vec{R}} + \frac{1}{2}\left(\sum_{\ell} m_\ell \dot{\vec{s}}_\ell\right)\cdot\dot{\vec{R}} \\
&\quad + \dot{\vec{R}}\cdot\frac{1}{2}\sum_{\ell} m_\ell \dot{\vec{s}}_\ell + \frac{1}{2}\sum_{\ell} m_\ell \dot{\vec{s}}_\ell \dot{\vec{s}}_\ell \\
&= \frac{1}{2}M\|\vec{V}\|^2 + 0 + 0 + \frac{1}{2}\sum_{\ell} m_\ell \|\dot{\vec{s}}_\ell\| \text{ by (14b)} \\
&= TC + T_c.
\end{aligned}
$$

where $T(C)$ is the kinetic energy of the center of mass and T_c is the kinetic energy of the system due to its motion relative to the center of mass.

Thus we have prove the following

Proposition 4 The kinetic energy of the system is the sum of the following two terms

(1) The kinetic energy of the center of mass $T(C)$.

(2) The kinetic energy T_c of the system due to its motion relative to the center of mass.

5.5   Conservation Principles

In **Chapter 3** we studied three conservation principles and two weaker versions of them, all for a single particle. Now we extend them for the system (1).

Proposition 5 (Principle Of Conservation Of Linear Momentum) If the sum of the external forces acting on the particles of the system (1) is equal to zero, then the linear momentum of the system remains constant along each of its trajectories.

Proof We consider the linear momentum $\vec{p}(t) = \sum_{\ell} \vec{p}_\ell(t)$ along a trajectory of the system and differentiate it with respect to time to get

$$
\begin{aligned}
\dot{\vec{p}}(t) &= \sum_\ell \dot{\vec{p}}_\ell(t) = \sum_\ell F_\ell \\
&= \sum_\ell F_\ell^{(e)} + \sum_{\ell,j(\ell \neq j)} F_{\ell j} \\
&= \sum_\ell F_\ell^{(e)} + 0, \text{ since } F_{\ell j} = -F_{j\ell} \\
&= 0.
\end{aligned}
$$

Thus we get $\frac{d\vec{p}(t)}{dt} \equiv 0$ along the trajectory. Hence the map $t \longmapsto \vec{p}(t)$ is constant along the trajectory.

This completes the proof.

In particular, for a *closed system*, we have $F_\ell^{(e)} = 0$ and hence $\sum_\ell F_\ell^{(e)} = 0$ and hence we have the following

Corollary 1 If the system is closed, then the linear momentum of the system is constant along any trajectory of the system.

Moreover, we have $\vec{p}(t) = M\vec{V}(t)$ by (9) and **Definition 2**. Consequently, if $\sum_\ell F_\ell^{(e)} = 0$ then $\vec{p}(t)$ is constant and so is $\vec{V}(t)$. Thus, we have the following corollary to **Proposition 5**.

Corollary 2 If sum $\sum_\ell F_\ell^{(e)} = 0$ then the center of mass of the particles either remains stationary or moves with constant velocity (along a straight line).

In many situations $\sum_\ell F_\ell^{(e)}$ is not identically zero, but there is a direction represented by a unit vector $\vec{e}$- along which $\sum_\ell F_\ell^{(e)}$ has no component. Then we prove the following

Proposition 6 If $\sum_\ell F_\ell^{(e)}$ is orthogonal to $\vec{e}$, then along each trajectory, the component of $\vec{p}(t)$ in the direction $\vec{e}$ remains constant.

Proof Considering the map $t \longmapsto \vec{p}(t) \cdot \vec{e}$ along a trajectory, we have to prove its constancy. Now we have

$$
\begin{aligned}
\frac{d}{dt}(\vec{p}(t) \cdot \vec{e}) &= \dot{\vec{p}}(t) \cdot \vec{e} = \left(\sum_\ell F_\ell^{(e)} \right) \cdot \vec{e} \\
&= 0 \text{ since } \sum_\ell F_\ell^{(e)} \perp \vec{e}.
\end{aligned}
$$

Hence the function $t \longmapsto \vec{p}(t) \cdot \vec{e}$ is constant along the trajectory; proving the proposition.

Proposition 7 (Conservation Principle Of Angular Momentum) Let A be a point in space (its position vector being $\vec{a}$). If the torque about the point A of the forces acting on the particles is identically zero, then the angular momentum of the system taken about A remains constant along each trajectory of the system.

Proof We have

$$\vec{L}_A(t) = \sum_\ell (\vec{r}_\ell - \vec{a}) \times \vec{p}_\ell = \sum_\ell (\vec{r}_\ell - \vec{a}) \times m_\ell \dot{\vec{r}}_\ell.$$

Therefore

$$
\begin{aligned}
\dot{\vec{L}}_A(t) &= \sum_\ell \dot{\vec{r}}_\ell \times m_\ell \dot{\vec{r}}_\ell + \sum_\ell (\vec{r}_\ell - \vec{a}) \times m_\ell \ddot{\vec{r}}_\ell \\
&= 0 + \sum_\ell (\vec{r}_\ell - \vec{a}) \times \vec{F}_\ell = \vec{N}_A(t) \\
&= 0 \text{ by assumption}
\end{aligned}
$$

Hence, along each trajectory, the map $t \longmapsto \vec{L}_A(t)$ is constant.

Again, in many situations, the torque N_A may not be identically zero. But there may exist a direction $\vec{e}$ along which N_A does not have component (that is, $\vec{N}_A \perp \vec{e}$). Then we have the following result.

Proposition 8 If $N_A \cdot \vec{e} \equiv 0$, then along any trajectory of the system, the function $t \longmapsto \vec{L}_A(t) \cdot \vec{e}$ remains constant.

Proof We have

$$\frac{d}{dt}(\vec{L}_A(t) \cdot \vec{e}) = \dot{\vec{L}}_A(t) \cdot \vec{e} = \vec{N}_A(t) \cdot \vec{e} = 0.$$

Hence the constancy of the map $\vec{L}_A(t) \cdot \vec{e}$ along any trajectory.

This completes the proof. □

Proposition 9 (Principle Of Conservation Of Total Energy) The total energy of a conservative system of particles along each of its trajectories remains constant.

Proof Let $t \longmapsto E(t)$ be the total energy function along an orbit. We have

$$
\begin{aligned}
\frac{d}{dt}(E(t)) &= \frac{d}{dt}\left(\frac{1}{2}\sum_\ell m_\ell \|\dot{\vec{r}}_\ell(t)\|^2 + U(\vec{r}(t), \vec{r_2}(t)\cdots, \vec{r}_n(t))\right) \\
&= \frac{1}{2}\sum_\ell m_\ell \frac{d}{dt}(\dot{\vec{r}}_\ell(t) \cdot \dot{\vec{r}}_\ell(t)) + \frac{d}{dt}U(\vec{r}_1(t), \vec{r_2}(t)\cdots, \vec{r}_n(t)) \\
&= \sum_\ell m_\ell \ddot{\vec{r}}_\ell(t) \cdot \dot{\vec{r}}_\ell(t) + \sum_\ell \text{grad}_\ell U(\vec{r}_1 \cdots \vec{r}_n) \cdot \dot{\vec{r}}_\ell(t) \\
&= \sum_\ell \left(m_\ell \ddot{\vec{r}}_\ell(t) + \text{grad}_\ell U\right) \cdot \dot{\vec{r}}_\ell(t) \\
&= \sum_\ell 0 \cdot \dot{\vec{r}}_\ell(t) \text{ since } m_\ell \ddot{\vec{r}}_\ell = -\text{grad}_\ell U.
\end{aligned}
$$

Hence the map $t \longmapsto E(t)$ is constant along each trajectory.

This completes the proof.   □

5.6   The Two Body Problem

In this article, we consider a mechanical system consisting of two particles P_1 and P_2 having masses m_1 and m_2. The forces acting on them have the following properties:

(a) There are no external forces.

(b) Internal forces are conservative.

(c) The potential energy function (of the internal forces) $U(\vec{r}_1, \vec{r}_2)$ depends only on the distance $\|\vec{r}_1 - \vec{r}_2\|$ between the particles $U(\vec{r}_1, \vec{r}_2) = W(\|\vec{r}_1 - \vec{r}_2\|)$ for some (smooth) function $W : (0, \infty) \longrightarrow \mathbb{R}$.

Such pairs of particles together with the forces of the type described above in (a), (b), (c) occurs in many situations. A pair of stars not far away from each other and hence exerting gravitational (attractive) forces on each other (but not colliding), remote enough from other heavenly bodies are studied extensively in Celestial Mechanics. Here $U(\vec{r}_1, \vec{r}_2) = \frac{-\gamma m_1 m_2}{\|\vec{r}_1 - \vec{r}_2\|}$ (and of course, $W(x) = \frac{-\gamma m_1 m_2}{x}$).

The study of motion of such a pair of particles (with the type of forces mentioned above) is known as the *two body problem.*

We analyse the two-body problem now. Without loss of generality, we assume that the motion is being studied from $t = 0$ onwards. Now, because no external forces are acting on the system, its center of mass either moves with constant velocity or is stationary. Suppose $\vec{a}$ is the position vector of it at $t = o$ and $\vec{b}$ its (constant) velocity. Then clearly, the trajectory of C (the center of mass) is given by $\vec{R}(t) = \vec{a} + t\vec{b}$.

At this stage, we introduce an auxiliary frame $\hat{\mathcal{F}}$ and use it to describe the motion of $\{P_1, P_2\}$ relative to the center of mass. The frame $\hat{\mathcal{F}}$ is specified by the following properties

(a) Origin of $\hat{\mathcal{F}}$ coincides with the center of mass for all $t \geq 0$.

(b) Axes of $\hat{\mathcal{F}}$ are parallel to the axes of $\mathcal{F}$.

Thus, $\hat{\mathcal{F}}$ is a frame moving with constant speed but without rotation relative to the stationary frame $\mathcal{F}$. The center of mass of $\{P_1, P_2\}$ is permanently at the origin of $\hat{\mathcal{F}}$. Consequently, the motion of $\{P_1, P_2\}$ relative to the center of mass is the same as motion relative to $\hat{\mathcal{F}}$.

Let $\vec{s}_1(t)$ be the position vector of P_1 with respect to $\hat{\mathcal{F}}$ and let $\vec{s}_2(t)$ be that of P_2. Clearly we have

$$\left.\begin{array}{rl} \vec{r}_1(t) = & \vec{a} + t\vec{b} + \vec{s}_1(t) \\ \vec{r}_2(t) = & \vec{a} + t\vec{b} + \vec{s}_2(t) \end{array}\right\} \tag{16}$$

The formulae (16) make it clear that it is enough to find the trajectories $t \longmapsto \vec{s}_1(t)$ and $t \longmapsto \vec{s}_2(t)$ of the particles relative to the moving frame $\hat{\mathcal{F}}$.

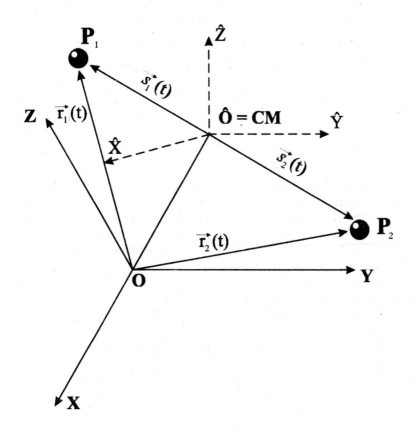

Figure 5.1

Note that $\vec{r}_1(t) - \vec{r}_2(t) = \vec{s}_1(t) - \vec{s}_2(t)$. We put it equal to $\vec{s}(t)$.
Now, equations of motion of the particles are

$$m_1 \ddot{\vec{r}}_1 = - \operatorname{grad}_{(\vec{r}_1)} W \left(\|\vec{r}_1 - \vec{r}_2\| \right)$$
$$m_2 \ddot{\vec{r}}_2 = - \operatorname{grad}_{(\vec{r}_2)} W \left(\|\vec{r}_1 - \vec{r}_2\| \right).$$

Note that

$$\operatorname{grad}_{(\vec{r}_1)} W \left(\|\vec{r}_1 - \vec{r}_2\| \right) = \operatorname{grad}_{\vec{s}} W \left(\|\vec{s}\| \right) \text{ and }$$

$$\operatorname{grad}_{(\vec{r}_2)} W \left(\|\vec{r}_1 - \vec{r}_2\| \right) = - \operatorname{grad}_{\vec{s}} W \left(\|\vec{s}\| \right).$$

Using these observations, we get

$$m_1 \ddot{\vec{r}}_1 = - \operatorname{grad}_{\vec{s}} W \left(\|\vec{s}\| \right)$$
$$m_2 \ddot{\vec{r}}_2 = + \operatorname{grad}_{\vec{s}} W \left(\|\vec{s}\| \right)$$

Multiplying these equations by m_1 and m_2 respectively, we get

$$m_1 m_2 \ddot{\vec{r}}_1 = -m_2 \, \mathrm{grad}_{\vec{s}} W \left(\|\vec{s}\| \right)$$
$$m_1 m_2 \ddot{\vec{r}}_2 = +m_1 \, \mathrm{grad}_{\vec{s}} W \left(\|\vec{s}\| \right)$$

Subtracting the second equation from the first, we get

$$m_1 m_2 (\ddot{\vec{r}}_1 - \ddot{\vec{r}}_2) = -(m_1 + m_2) \, \mathrm{grad}_{\vec{s}} W \left(\|\vec{s}\| \right).$$

Using $\vec{s} = \vec{s}_1 - \vec{s}_2 = \vec{r}_1 - \vec{r}_2$, the above equation becomes:

$$m_1 m_2 \ddot{\vec{s}} = -(m_1 + m_2) \, \mathrm{grad}_{\vec{s}} W (\|\vec{s}\|)$$

or equivalently

$$\frac{m_1 m_2}{(m_1 + m_2)} \ddot{\vec{s}} = - \, \mathrm{grad}_{\vec{s}} W (\|\vec{s}\|) \qquad (17)$$

Clearly, the last equation is the equation of motion of a particle. Indeed we consider a particle, we call it Q, having mass $\frac{m_1 m_2}{m_1 + m_2}$, with its position vector $\vec{s}(t)$. Then equation (17) is precisely the equation of motion of the particle Q, it being acted upon by a conservative force having potential energy function $\vec{s} \longmapsto W(\|\vec{s}\|)$.

Thus, with the pair $\{P_1, P_2\}$ of particles, we have associated a third hypothetical particle Q moving in a central force field. We studied such motion in detail in **Chapters 1, 2, 3 and 4**. We suppose therefore that we have solved the problem of motion of the particle Q and obtained its trajectory $t \longmapsto \vec{s}(t)$.

We prove below that the motion of P_1 and P_2 is expressible in terms of the motion of Q.

Note first the identity: $m_1 \vec{s}_1(t) + m_2 \vec{s}_2(t) \equiv 0$. This identity follows from the fact that the center of mass of the pair $\{P_1, P_2\}$ in the frame $\hat{\mathcal{F}}$ is given by $\frac{m_1 \vec{s}_1(t) + m_2 \vec{s}_2(t)}{m_1 + m_2}$. On the other hand, we have chosen $\hat{\mathcal{F}}$ in such a way that the center of mass is always at its origin and therefore $\frac{m_1 \vec{s}_1(t) + m_2 \vec{s}_2(t)}{m_1 + m_2} = 0$ which implies the above identity. $m_1 \vec{s}_1(t) + m_2 \vec{s}_2(t) \equiv 0$.

Now we consider the following simultaneous equations in $\vec{s}_1(t)$ and $\vec{s}_2(t)$:

$$m_1 \vec{s}_1(t) + m_2 \vec{s}_2(t) = 0 \ and \ \vec{s}_1(t) - \vec{s}_2(t) = \vec{s}(t).$$

Solving them we get $\vec{s}_1(t)$ and $\vec{s}(t)$ gives us $\vec{s}_1(t) = \frac{m_2 \vec{s}(t)}{m_1 + m_2}$ and $\vec{s}_2(t) = \frac{-m_1 \vec{s}(t)}{m_1 + m_2}$. This pair of equations combined with equations (16) give us the trajectories of the particles P_1 and P_2: in $\mathcal{F}$

$$\vec{r}_1(t) = \vec{a} + t\vec{b} + \frac{m_2 \cdot \vec{s}(t)}{m_1 + m_2}$$
$$\vec{r}_2(t) = \vec{a} + t\vec{b} - \frac{m_1 \vec{s}(t)}{m_1 + m_2}$$

With this, we have justified our claim made earlier that the motion of the system $\{P_1, P_2\}$ can be studied in terms of the single hypothetical particle Q.

5.7 Some Solved Examples

Example 1 A particle P having mass m, speed v collides with a stationary particle Q having mass M. Assuming that no energy is lost in the collision, show that the maximum speed which the particle Q can acquire is $u = \frac{2mv}{m+M}$.

Solution We make use of the two conservation principles (i) total linear momentum and (ii) total energy. Since there are no forces acting on the particles, the total energy is the sum of the kinetic energies of the particles.

Let w denote the speed of the particle P after the collision. Clearly, the velocities $\vec{u}, \vec{v}$, and $\vec{w}$ are all in the same direction. Now the principle (i) gives

$$mv = mw + Mu \tag{$*$}$$

and so, $w = \frac{mv - Mu}{m}$. The principle (ii) gives the equation

$$\frac{m}{2}v^2 = \frac{m}{2}w^2 + \frac{M}{2}u^2 \tag{$**$}$$

Substituting $w = \frac{mv - Mu}{m}$ in ($**$) we get

$$mv^2 = m\left(\frac{mv - Mu}{m}\right)^2 + Mu^2$$

Therefore, $m^2v^2 = m^2v^2 - 2Mmuv + M^2u^2 + mMu^2$ which gives

$$2Mmuv = M(M + m)u^2.$$

Therefore, either $u = o$ or $u = \frac{2mv}{(M+m)}$.

Thus the maximum speed Q can acquire is $\frac{2mv}{(M+m)}$. $\qquad\square$

Example 2 Particles $P_1, P_2, \cdots, P_n$ move in a plane. Prove

(i) At any instant t during the motion, there is a straight line $W(t)$ such that the total angular momentum of the system taken about any point of $W(t)$ is zero.

(ii) If no forces are acting on the particles then the line does not change with time.

Solution We consider an arbitrary instant t_o of time. Having chosen the instant t_o, we simplify our treatment by not mentioning it. Thus, we write $\vec{r}$ for $\vec{r}(t_o)$, $\vec{L}$ for $\vec{L}(t_o)$ and so on. Without loss of generality, we make the following simplifying assumptions.

(a) The plane of motion of the particles is the XOY-plane.

(b) The origin of the frame of reference is chosen in such a way that $\vec{L}(= \vec{L}(t_o))$ is a non-zero vector. (See Exercise No. 7(b) at the end of this chapter.)

Note that the assumption (a) implies that the (non-zero) $\vec{L}_o$ is along the z-axis of the chosen frame of reference.

Now we consider the velocity $\vec{V}(= \vec{V}(t_o))$ of the center of mass.

Suppose, first that $\vec{v} \neq 0$.

Next, we consider a point A having position vector $\vec{a}(= \vec{a}(t_o))$ such that the angular momentum of the system about A is zero:

$$\sum_{\ell} (\vec{r}_\ell - \vec{a}) \times m_\ell \dot{\vec{r}}_\ell = 0$$

and hence,

$$\sum m_\ell \vec{r}_\ell \times \dot{\vec{r}}_\ell = \vec{a} \times \sum m_\ell \dot{\vec{r}}_\ell.$$

that is

$$\vec{L}_o = M \vec{a} \times \vec{V} \qquad\qquad (*)$$

Conversely, it is clear that if a vector $\vec{a}$ satisfies $(*)$ then the angular momentum of the system taken about $\vec{a}$ is zero. Consequently, (under the on going assumption $\vec{V} \neq 0$) we have to prove that the set of all points A having position vector $\vec{a}(= \vec{a}(t_o))$ satisfying $(*)$ contains a straight line.

To see this, choose two unit vectors $\vec{e}$, and $\vec{f}$ in the XOY-plane such that (i)$\vec{e}$ and $\vec{f}$ are perpendicular to each other, (ii) $\vec{f}$ is parallel to $\vec{V}$ (and therefore $\vec{V} = V \cdot \vec{f}$).

Now, we consider the straight line

$$W = \left\{ \frac{L_o}{MV} \vec{e} + s\vec{f}, s \in \mathbb{R} \right\}$$

Clearly, for any point $\frac{L_o}{MV} \vec{e} + s\vec{f}$ we have

$$\left(\frac{L_o}{MV} \vec{e} + s\vec{f} \right) \times MV\vec{f} = L_o \vec{e} \times \vec{f} + 0 = L_o \vec{k} = \vec{L}_o.$$

In other words, every point of W satisfies $(*)$. Hence every point of W has the property that the angular momentum of the system about it is zero.

Note that if there are no forces acting on the system then $\vec{L}(t_o), \vec{V}(t)$, $\vec{e}, \vec{f}$ are all independent of time and hence the line W does not change with time.

Lastly, we consider the case in which $\vec{V} = 0$.

Now we consider a frame $\hat{\mathcal{F}}$ which is moving with uniform velocity relative to $\mathcal{F}$. In the frame $\hat{\mathcal{F}}$, the center of mass will have non-zero velocity and the above discussed method will provide a straight line $\hat{W}$ with the desired property. This straight line $\tilde{W}$ is a (moving) straight line W in $\mathcal{F}$ with the desired property.   □

Example 3 Show that, three particles having equal mass m which are acted upon by their gravitational forces can be located at three points of a straight line, at an equal distance a apart provided that the line rotates about an axis through the central point with angular velocity $\sqrt{\left(\frac{5\gamma m}{4a^3}\right)}, \gamma$ being the universal gravitational constant.

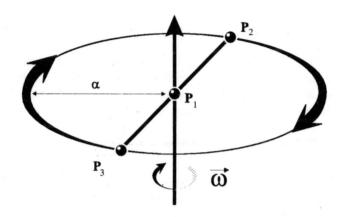

Figure 5.2

Solution The middle particle P_1 is acted upon by equal and opposite forces of attraction, (from the particles P_2 and P_3) and therefore, there is no net force acting on it.

It is clear that the forces acting on P_2 and P_3 are equal and opposite. We therefore consider the motion of one of them, say P_3.

There are two forces acting on P_3: (i) gravitational attraction due to P_1 having magnitude $\frac{\gamma m^2}{a^2}$ and (ii) that due to P_3 having magnitude $\frac{\gamma m^2}{4a^2}$, both these forces being along $\vec{P_3 P_2}$. Therefore, the total force acting on P_3 is $\frac{\gamma m^2}{a^2} + \frac{\gamma m^2}{4a^2} = \frac{5\gamma m^2}{4a^2}$, the direction of the force being from P_3 towards P_2. Now when P_3 is rotating about P_1 with angular velocity w, when at a distance a from it, it is acted upon by a centrifugal force of magnitude mw^2a^2 in the outward direction. Clearly, the particle P_3 will remain at the same distance a from P_1 if

$$\frac{5\gamma m^2}{4a^2} = mw^2a.$$

This gives $w = \sqrt{\left(\frac{5\gamma m}{4a^3}\right)}$.   □

Example 4 Two gravitating masses m_1 and $m_2 (m_1 + m_2 = M)$ are separated at a distance a and then are released from rest to gravitate towards each other. Show that when the separation is $r(r < a)$, the velocities of the particles are:

$$v_1 = m_2\sqrt{\frac{2\gamma}{M}\left(\frac{1}{r} - \frac{1}{a}\right)}, \text{ and } v_2 = m_1\sqrt{\frac{2\gamma}{M}\left(\frac{1}{r} - \frac{1}{a}\right)}.$$

Solution Consider the single particle Q having mass $m = \frac{m_1 m_2}{M}$. The equation of motion of the particle is $\frac{m_1 m_2}{M}\ddot{\vec{r}}(t) = \frac{-m_1 m_2 \gamma}{r^2}\vec{e}_r(t)$, that is

$$\frac{\ddot{\vec{r}}}{M}(t) = \frac{-\gamma}{r^2}\vec{e}_r(t) \tag{$*$}$$

Note that the distance r between the gravitating particles P_1, P_2 is also the distance of the particle Q from the center of mass of $\{P_1, P_2\}$. Also, note that after the particles are released from rest, the motion becomes one dimensional and hence the equation of motion ($*$) takes the form

$$\ddot{r}(t) = -\frac{\gamma M}{r^2(t)} \tag{$**$}$$

Multiplying it by $2\dot{r}$ and integrating, we get $\frac{\dot{r}^2}{M} = \frac{2\gamma}{r} + C$. The initial condition $\dot{r} = 0$ when $r = a$ implies $C = \frac{-2\gamma}{a}$. Therefore

$$\dot{r} = \sqrt{2M\gamma\left(\frac{1}{r} - \frac{1}{a}\right)}$$

Also, we have $m_1 r_1 + m_2 r_2 = 0$ and $r_1 - r_2 = r$. (As in case of the two-body problem). These equations imply $m_1 v_1 + m_2 v_2 = 0$ and $v_1 - v_2 = \dot{r}$. Solving, these equations for v_1 and v_2, we get

$$v_1 = m_2\sqrt{\frac{2\gamma}{M}\left(\frac{1}{r} - \frac{1}{a}\right)} \text{ and } v_2 = m_1\sqrt{\frac{2\gamma}{M}\left(\frac{1}{r} - \frac{1}{a}\right)}$$

□

EXERCISES

1. Show that if the total external force acting on a system of particles is zero, then the total torque taken about any point has the same value.

2. A system consists of three particles having one, two and three units of mass respectively. At time $t = 0$, the positions and velocities of the particles are

$$\vec{r}_1 = \vec{i} + \vec{j}, \quad \vec{r}_2 = \vec{j} + \vec{k}, \quad \vec{r}_3 = \vec{k}$$
$$\vec{v}_1 = 2\vec{i}, \quad \vec{v}_2 = \vec{j}, \quad \vec{v}_3 = \vec{i} + \vec{j} + \vec{k}.$$

(a) Find the linear momentum of the system.

(b) Find the angular momentum of the system (i) about the origin, (ii) about the point $\vec{i} + 2\vec{j} + 3\vec{k}$.

3. Three particles of masses 2, 3 and 5 units respectively move under the influence of a force field such that their position vectors are given by

$$
\begin{aligned}
\vec{r}_1(t) &= 2\vec{i} + 3\vec{j} + t^2\vec{k}, \\
\vec{r}_2(t) &= (t+1)\vec{i} + 3t\vec{j} - 4\vec{k} \text{ and} \\
\vec{r}_3(t) &= t^2\vec{i} - t\vec{j} + (2t-1)\vec{k}.
\end{aligned}
$$

(a) Find the angular momentum and the torque of the system at the time $t = 2$.

(b) Total angular momentum and total torque of the system about the point $\vec{a} = \vec{i} - 2\vec{j} + 3\vec{k}$ when $t = 1$.

4. Show that the kinetic energy relative to the center of mass of a system of particles may be written in the form

$$
\frac{1}{2} \sum_{i>j} \frac{(m_i m_j v_{ij}^2)}{M}
$$

where m_ℓ is the mass of the particle P_ℓ, $\sum_\ell m_\ell = M$ and v_{ij} is the speed of P_j with respect to P_i.

5. Two particles, each of mass m are connected by a rigid mass-less rod of length a. The system is initially at rest on a smooth horizontal table. At time $t = 0$, a force F begins to act on one of the particles. The magnitude of the force remains constant while the direction of the force is normal to the rod all the time. Determine by using work energy relation, the angular velocity of the rod at a time when the rod makes an angle θ with its original position.

6. Two planets of masses m_1 and m_2 describe ellipses with the sun at their focii, their major axes being $2a_1$ and $2a_2$. Show that their periods satisfy $\frac{T_1}{T_2} = \frac{a_1^3(M+m_2)}{a_2^3(M+m_1)}$, where M is the mass of the sun.

7. For a system of particles $\{P_1 \cdots P_n\}$ and for any two points A and B with their position vectors $\vec{a}$ and $\vec{b}$ respectively, prove

(a) $\vec{L}_A = \vec{L}_B + (\vec{a} - \vec{b}) \times M\vec{V}$, where $\vec{V}$ is the velocity of the centre of mass, of the system.

(b) Use result (a) to prove that when the center of mass is not stationary, we can choose a point A in space in such a way that $\vec{L}_A \neq 0$.

8. Two particles of masses $4m$ and m are free to move along the X-axis. There is a force of attraction between the two particles having magnitude kr, where k is a constant and r is the distance between them. At time $t = 0$ the particle having mass $4m$ is located at $x = 5a$ and the other particle is at $x = 10a$. Both of them are at rest (when $t = 0$).

    (a) At what value of x do the two particles collide?

    (b) What is the relative velocity of the particles when they collide?

9. Two particles A and B of masses m and M respectively, describe orbits under their mutual attraction. Show that the total angular momentum of the system with respect to the center of mass is given by $\frac{M\vec{\ell}}{M+m}$ where $\vec{\ell}$ is the angular momentum of the particle A with respect to the position of B.

10. Two particles of masses m and M are connected by an elastic string having natural length a and spring constant k. Initially, they are at rest on a horizontal table at a distance a apart. The mass m is set in motion with a speed v in a direction perpendicular to the string. Prove that in the subsequent motion before the spring becomes slack, the greatest distance r between the particles is a root of the equation

$$m \cdot M \cdot v^2 \cdot (r + a) = k(m + M)r^2(r - a).$$

11. The only forces acting on two particles P_1, P_2 each having mass m is their mutual attraction having magnitude $\frac{\mu m}{r^2}$. Initially, the particles are at a distance $2a$ apart and are moving with velocities

$$\frac{1}{2}\sqrt{(\frac{\mu}{a})} \cdot (\sqrt{3}\vec{i} + \vec{j}) \text{ and } \frac{1}{2}\sqrt{(\frac{3\mu}{a})}(-\vec{i} + \sqrt{3}\vec{j})$$

    where $\vec{i}$ is a unit vector in the direction $P_1 P_2$ and $\vec{j}$ is a unit vector perpendicular to $\vec{i}$. Show that their center of mass moves with constant velocity $\sqrt{(\frac{\mu}{a})}\ \vec{j}$

12. Two particles moving under their mutual gravitational attraction describe circular orbits about one another with a period T. If they are suddenly stopped in their orbits and then allowed to gravitate towards each other, show that they will collide after a time $\frac{T}{4\sqrt{2}}$.

13. For a system of n particles, show that if the total angular momentum about the center of mass is non-zero at some time, then it is not possible for all the particles to come together in the subsequent motion.

14. Show that three particles under their mutual gravitational attraction can be located at the vertices of an equilateral triangle with sides of constant length a provided that the triangle rotates about the axis through the center of mass perpendicular to its plane with angular velocity $\sqrt{\frac{\gamma(m_1+m_2+m_3)}{a^3}}$.

15. Three starts each of mass M are acted upon by their mutual gravitational attraction. Prove that if the starts are moving with constant angular velocity w, placed at equal distances on a circle of radius a, then $w^2 = \frac{\gamma M}{\sqrt{3} \cdot a^3}$

Chapter 6

Rigid Dynamics

"The reader may need an effort of will to perceive mathematics as a tutor of our spatial imagination. More frequently, one associates mathematics with rigorous logic and computational formalism. But this is only discipline, a ruler with which we are taught not to die".

Yu. I. Manin

6.1 Introduction

Untill now, we have considered the motion of families of particles; these so called *particles* being the mathematically idealizations of small bits of matter. Naturally then, the next step is to consider a more realistic model of a piece of matter to which the laws of particle mechanics could be extended by means of the techniques of integral calculus. We therefore consider a solid non-deformable piece of matter having moderate size, which we have been calling a **rigid body**.

We know that every piece of matter is granular in constitution. Thus, a rigid body is a large assembly of its constituent grains namely, its molecules. These grains of matter are so minute that they behave perfectly like the particles. Consequently, a rigid body is a family of particles

$$\{P_1, P_2, \cdots, P_n\} \tag{1}$$

having their respective masses $\{m_1, m_2, \cdots, m_n\}$. The rigidity of the body now means that the distance between any two of its particles remains constant:

$$\text{dist } \{P_i, P_j\} = \text{ constant } ; 1 \leq i, j \leq n.$$

But, we should bear in our minds that though we are considering a rigid body as a family of particles, we are interested in its motion as a single entity and not in the movements of its individual particles.

On the other hand, the grains of matter within the body are so densely populated that (and also, having decided not to pay attention to the individual particles) it is convenient to use the statistical averages and treat the discrete assembly (1) as if it were a continuous distribution of matter having definite, non-deforming shape and hence a constant volume also.

In fact, in practice, rigid bodies are given more often in the form of continuous distribution of matter, than a discrete assemblies of particles. But the basic laws of motion are applicable only to particle families. Therefore, to apply laws of motion to such a continuous model, we consider a conceptual division of the body into a finite number of its bits $\{B_1, B_2, \cdots, B_n\}$, each bit B_i being so small that it can be regarded as a particle to which Newton's equations of motion are applicable. Thus Newton's equations of motion are made applicable to the conceptual family $\{B_1, B_2, \cdots, B_n\}$ of particles. Combining the resulting information about the motion of the B_j^s, we get an approximation to the motion of the body. Naturally, this approximation improves as we go on refining the division. Consequently, the motion of a rigid body itself is the limiting case of the motion of the family $\{B_1, B_2, \cdots, B_n\}$ as we go on refining the division.

Actually, this heuristic consideration gives results which are in perfect agreement with the actually observed motion.

We therefore consider both the models of a rigid body. To be more precise, we have the following two models of a rigid body.

(I) The Discrete Model A rigid body is a finite (but arbitrarily large) assembly (1) of particles which move in such a way that the distances between the particles: dist $\{P_i, P_j, \} 1 \leq i, j \leq n$, remains constant.

(II) The Continuous Model A rigid body is a continuous distribution of matter having a mass density ρ and a shape B which is not deformable, the volume of B being finite.

We denote the discrete model of a rigid body by

$$\left\{ \begin{matrix} P_1, & P_2, & \cdots & P_n \\ m_1, & m_2, & \cdots & m_n \end{matrix} \right\} \tag{2}$$

and the continuous model by

$$\{B, \rho\} \tag{3}$$

Throughout this chapter, we will assume that in case of the discrete model, the assemby (1) has at least three non-collinear points. Moreover if we are considering the rotational motion of the rigid body about a fix point, we will also assume that none of the three particles is at the center of rotation.

As noted above, we come across the continuous model in practice while the basic laws of motion are directly applicable to the discrete model. Consequently, it is necessary to keep both models in our minds. Actually, we will derive all the results of rigid dynamics first in case of discrete model and then extend them to the continuous model. This procedure of extension from discrete to continuous model is accomplished by means of the familiar integral calculus.

We conclude this section by discussing an example in which we calculate the kinetic energy of a thin rod (continuous model) by treating it as an approximating discrete assembly of its tiny bits.

Example 1 A uniform thin rod rod has mass M. Its two end points A and B are moving with velocities $\vec{u}$ and $\vec{v}$ respectively. Prove that its kinetics energy T is given by

$$T = \frac{M}{6}[\vec{u} \cdot \vec{u} + \vec{u} \cdot \vec{v} + \vec{v} \cdot \vec{v}].$$

Solution Let ℓ be the length of the rod. Consider a point C on the rod which is at a distance x from the end A. Denoting by $\vec{a}(t)$ the position vector of the end A, by $\vec{b}(t)$ that of the end B and by $\vec{r}(t)$ that of the point C, we have the equation

$$\vec{r}(t) = \left(1 - \frac{x}{\ell}\right)\vec{a}(t) + (\frac{x}{\ell})\vec{b}(t).$$

Differenting this equation, we get

$$\dot{\vec{r}}(t) = \left(1 - \frac{x}{\ell}\right)\dot{\vec{a}}(t) + (\frac{x}{\ell})\dot{\vec{b}}(t).$$

But, $\dot{\vec{a}}(t) = \vec{u}$ and $\dot{\vec{b}}(t) = \vec{v}$ (given). Hence the instantaneous velocity $\vec{v}(t)$ of the point C is given by

$$\vec{v}(t) = \left(1 - \frac{x}{\ell}\right)\vec{u} + \frac{x}{\ell}\vec{v}. \tag{$*$}$$

Next, we consider the division of the rod into tiny bits $B_1, B_2, \cdots B_n$ as shown in Fig 6.1. Let $\delta x_1, \delta x_2, \cdots \delta x_n$ be their respective lengths. We also choose points: ζ_1 in B_1, ζ_2 in $B_2 \cdots \zeta_n$ in B_n the choice of point being arbitrary.

We consider each piece B_j as a particle situated at the point ζ_j. Clearly its mass is proportional to its length and hence equal to $\frac{M\delta x_j}{\ell}$. Its velocity will be $\vec{v}(\zeta_j) = \left(1 - \frac{\zeta_j}{\ell}\right)\vec{u} + \frac{\zeta_j}{\ell}\vec{v}$. according to $(*)$. Consequently its contribution δT_j to the kinetic energy of the rod is

$$\delta T_j = \frac{1}{2}\frac{M}{\ell}\delta x_j\|\vec{v}(\zeta_j)\|^2 = \frac{1}{2}\frac{M}{\ell}\|\left(1 - \frac{\zeta_i}{\ell}\right)\vec{u} + \frac{\zeta_j}{\ell}\vec{v}\|^2\delta x_j.$$

Hence the approximate kinetic energy of the rod is the sum

$$\sum_{j=1}^{n}\delta T_j = \frac{M}{2\ell}\sum_{j=1}^{n}\|\left(1 - \frac{\zeta_j}{\ell}\right)\vec{u} + \frac{\zeta_j}{\ell}\vec{v}\|^2\delta x_j.$$

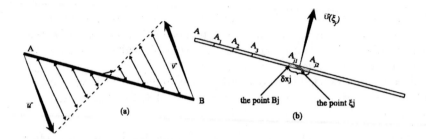

Figure 6.1

These approximations converge to the integral

$$\frac{M}{\ell}\int_0^1 \| \left(1 - \frac{x}{\ell}\right)\vec{u} + \frac{x}{\ell}\vec{v}\|^2 dx = \frac{M}{6}\left[\vec{u}\cdot\vec{u} + \vec{u}\cdot\vec{v} + \vec{v}\cdot\vec{v}\right]$$

as the division of the rod into the pieces $B_1 B_2 \cdots B_n$ refines, each piece B_j shrinking to a point.   □

6.2   The Nature of Motion

Now, we consider the problem of describing mathematically the instantaneous positions of a rigid body and allied dynamical observable (such as velocity, acceleration etc). Moreover, we want to adapt a model for the motion of the rigid body.

As already explained, a rigid body is basically an assembly of particles. It is therefore tempting to describe its instantaneous position in terms of the complete data consisting of the positions of the constituent particles at the instant under consideration. In the same vein, we may consider the motion of the body as the bundle of all the trajectories of the constituent particles.

Unfortunately, this collective way of describing the dynamics of a rigid body proves to be mathematically unwieldy. This is indeed so for the following reasons:

(a) The number of particles in a rigid body is frighteningly large, making it impossible to perceive the instantaneous positions and trajectories of all of them.

(b) Our interest is in the motion of the body as a single whole and not in that of its individual particles.

(c) The rigidity of the body which gives to it its shape, size, the mass distribution etc. has much influence on the motion. But the above collective model neglects the rigidity aspect altogether.

It is actually the rigidity of the body which gives rise to a more elegant model to which mathematical methods can be applied more effectively. This point is explained below.

To begin with we choose a reference point G in the body. For the time being, the choice of G is arbitrary but at a latter stage we will be specific. Now the instantaneous position of the body is uniquely determined by the following two things:

(1) The instantaneous position of the reference point G.

(2) The orientation of the body about G (in reference to the stationary surroundings.)

Consequently the motion of the body is determined by the following: (i) the displacement of the reference point G which is accompanied by (ii) a change occurring with time in the orientation of the body.

Of course, the effects of the rotational motion will be felt differently at different reference point G. But an important fact of rigid dynamics is that at each instant t, there exists a unique vector which determines rotational state of the body about any point G of it. This unique vector is the **instantaneous angular velocity** $\vec{w}(t)$ of the body. We will prove its existence and uniqueness in the next section.

To analyse the motion, we choose a stationary frame of reference $\mathcal{F}$ as follows: If the body is anchored at a point (and is kept freely rotating about the anchorage point) then the (stationary) $\mathcal{F}$ is chosen, so that its origin coincides with the point of anchorage. If the body is moving freely we choose any stationary $\mathcal{F}$.

We need an auxiliary frame of reference to describe the orientation of the body. So we choose another frame $\hat{\mathcal{F}}$ which is required to satisfy the following two conditions :(a) Origin of $\hat{\mathcal{F}}$ coincides with the chosen reference point G and (b) the body is at rest with respect to $\hat{\mathcal{F}}$ (and hence, $\hat{\mathcal{F}}$ is moves along with the body.)

We call such a frame $\hat{\mathcal{F}}$ a **body frame of reference**. (See fig 6.2)

Clearly, the motion of the body can be studied in terms of the motion of the frame $\hat{\mathcal{F}}$ relative to $\mathcal{F}$. In fact, at any instant, the location of the body is uniquely determined by (i) the instantaneous position of the reference point G and (ii) the orientation of the body about G. Both these items can be expressed more conveniently in terms of the positions of $\hat{\mathcal{F}}$ relative to $\mathcal{F}$. For, the instantaneous position of G is the instantaneous position of the origin of $\hat{\mathcal{F}}$ and the orientation of the body is the orientation of $\hat{\mathcal{F}}$(which can be expressed *mathematically* in terms of, say, the direction cosines of the axes of $\hat{\mathcal{F}}$ relative to the frame $\mathcal{F}$.)

We take unit vectors $\vec{u_1}, \vec{u_2}, \vec{u_3}$ along the axes of the body frame $\hat{\mathcal{F}}$. Clearly, $(\vec{u_1}, \vec{u_2}, \vec{u_3})$ is a triple of constant vectors with respect to $\hat{\mathcal{F}}$, but they form a time-varying right handed triple $(\vec{u_1}(t), \vec{u_2}(t), \vec{u_3}(t))$ of mutually perpendicular unit vectors when seen from the stationary frame $\mathcal{F}$.

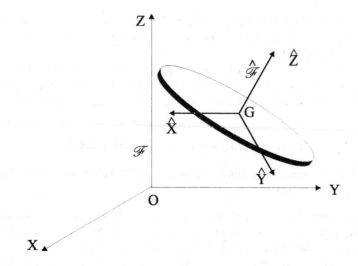

Figure 6.2

Now, we have two frames of reference: The stationary frame $\mathcal{F}$ and the moving body frame $\hat{\mathcal{F}}$. As usual, the equations of motion of the body must be formulated in reference to the stationary frame $\mathcal{F}$. On the other hand, because the body is stationary in $\hat{\mathcal{F}}$, many of the dynamical attributes of the body remain constant in the body frame $\hat{\mathcal{F}}$. On account of this simple fact, it is found more convenient to work in the body frame $\hat{\mathcal{F}}$. In particular, we will transfer the equation of motion of the body from the stationary frame $\mathcal{F}$ to the body frame $\hat{\mathcal{F}}$; which we will see then takes the form similar to $m\vec{a} = F$.

6.3   The Instantaneous Angular Velocity

In order to motivate the notion of instantaneous angular velocity, we consider the following simple rotational motion of a rigid body (say a spinning top.) Suppose, one point of the body is held fix. Also, suppose that there is a fix line W passing through the fixed point, this line W having the following property:

Each point P of the body appears to be revolving in a circular orbit with W as the axis with the same angular speed (that is, each point P is describing the same angle, w radians in one second while moving along its own circular orbit.)

We consider the unit vector $\vec{e}$ along W in such a way that all particles of the body are moving counter clockwise around $\vec{e}$. We form the vector $w \cdot \vec{e}$ and denote it by $\vec{w}$.

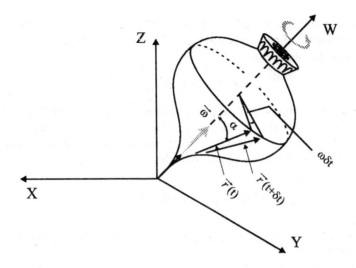

Figure 6.3

Now, we consider the motion of a point P of the body. Let $\vec{r}(t)$ denote its position vector at the instant t. Let α be the angle between the line W and the position vector $\vec{r}(t)$. The assumption that P is performing circular motion about the line W as its axis now means that the angle α does not change with time. We also consider the position vector $\vec{r}(t + \delta t)$ of the point P at the time $(t + \delta t)$. We denote the displacement $\vec{r}(t + \delta t) - \vec{r}(t)$ undergone by P in the interval $[t, t + \delta t]$. by $\delta \vec{r}(t)$. Then, the magnitude of the displacement $\delta \vec{r}(t)$ to within first order of smallness is given by $\delta r(t) = r \sin \alpha \cdot w \cdot \delta t$. Consequently the displacement vector $\delta \vec{r}(t)$ (to within first order of smallness) is given by

$$\delta \vec{r}(t) \simeq \vec{w}(t) \times \vec{r}(t) \delta t.$$

Dividing both sides of this equation by δt and taking limit as $\delta t \longrightarrow 0$ we get $\vec{v}(t) = lim_{\delta t \to 0} \frac{\delta \vec{r}(t)}{\delta t} = lim_{\delta t \to 0} \frac{\vec{w} \times \vec{r}(t) \delta t}{\delta t} = \vec{w} \times \vec{r}(t)$. Thus, we have proved that the velocity of the particle P is given by

$$\dot{\vec{r}}(t) = \vec{w} \times \vec{r}(t) \qquad (4)$$

Regard w (the common angular speed of the particles of the body) as the angular speed of the body, the vector $\vec{w} = w\vec{e}$ as its angular velocity and the line W as its axis of rotation. Equation (4) relates the velocity of a point P of the body to the angular velocity $\vec{w}$ of the body (using the position vector $\vec{r}(t)$. of P as a factor).

Actually equation (4) continuous to hold in the more general time-varying situation. More specifically, we will prove the following result : Suppose, the rigid body is rotating arbitrary (but in a smooth manner) about a fix point 0. Then at each instant t, there exists (i) a unique line

$W(t)$ passing through 0 and (ii) a vector $\vec{w}(t)$ along $W(t)$ in such a way that each point P of the body having position vector $\vec{r}(t)$ has the velocity $\dot{\vec{r}}(t)$ given by the equation

$$\dot{\vec{r}}(t) = \vec{w}(t) \times \vec{r}(t) \tag{5}$$

and consequently, the point P appears to be rotating *at that instant t* about the line $W(t)$ in a circular orbit with the instantaneous angular speed $w(t) = \|\vec{w}(t)\|$. Moreover, the vector $\vec{w}(t)$ and also the line $W(t)$ are unique. To prove this statement, we begin with the vector $\vec{w}(t)$ given by

$$\vec{w}(t) = \frac{1}{2} \sum_{j=1}^{3} \dot{\vec{u}}_j(t) \times \vec{u}_j(t).$$

where $\vec{u}_1(t), \vec{u}_2(t), \vec{u}_3(t)$ are unit vectors (introduced at the end of 6.2) along the axes of $\hat{\mathcal{F}}$.

We prove below that $\vec{w}(t)$ defined as above, satisfies equation (5). Towards this end, let P be any point of the body having position vector $\vec{r}(t)$. Resolving it with respect to the $\mathcal{F}$, we get

$$\vec{r}(t) = \sum_{j=1}^{3} (\vec{r}(t) \cdot \vec{u}_j(t)) \, \vec{u}_j(t). \tag{6}$$

Now, because P is stationary with respect to $\hat{\mathcal{F}}$, the coordinates of P (with respect to $\hat{\mathcal{F}}$) namely

$$((\vec{r}(t) \cdot \vec{u}_1(t)), \ (\vec{r}(t) \cdot \vec{u}_2(t)), \ (\vec{r}(t) \cdot \vec{u}_3(t))$$

do not change with time and therefore, we have

$$\frac{d}{dt}(\vec{r}(t) \cdot \vec{u}_j(t)) = 0 \qquad 1 \le j \le 3 \tag{7}$$

Consequently, we get

$$\dot{\vec{r}}_j(t) = \sum_{j=1}^{3} (\vec{r_j}(t) \cdot \vec{u}_j(t)) \dot{\vec{u}}_j(t) \tag{8}$$

Also, resolving $\dot{\vec{r}}(t)$ in the frame $\hat{\mathcal{F}}$, we get

$$\dot{\vec{r}}(t) = \sum_{j=1}^{3} (\dot{\vec{r}}(t) \cdot \vec{u}_j(t)) \vec{u}_j(t) \tag{9}$$

But, equation (7) gives $\dot{\vec{r}}_j(t)\vec{u}_j(t) + \vec{r_j}(t) \cdot \dot{\vec{u}}_j(t) = 0$. Therefore, from equation (9), we get

$$\dot{\vec{r}}(t) = -\sum_{j=1}^{3} (\vec{r}(t) \cdot \dot{\vec{u}}_j(t)) \vec{u}_j(t) \tag{10}$$

Adding (8) and (10), we get

$$2\dot{\vec{r}}(t) = \sum_{j=1}^{3} \vec{r}(t) \cdot \dot{\vec{u}}_j(t) - \sum_{j=1}^{3} (\vec{r}(t)) \cdot \dot{\vec{u}}_j(t))\vec{u}_j(t)$$

$$= \left(\sum_{j=1}^{3} \dot{\vec{u}}_j(t) \times \vec{u}_j(t) \right) \times \vec{r}(t).$$

Thus, putting $\vec{w}(t) = \sum_{j=1}^{3} \dot{\vec{u}}_j(t) \times \vec{u}_j(t)$ in the above equation, we get equation (5).

To prove the uniqueness part of the vector $\vec{w}(t)$ we consider two vectors $\vec{w}_1(t)$ and $\vec{w}_2(t)$ both satisfying equation (5):

$$\dot{\vec{r}}(t) = \vec{w}_1(t) \times \vec{r}(t) = \vec{w}_2(t) \times \vec{r}(t).$$

for all the points $P(= \vec{r}(t))$ of the body. This gives us:

$$[\vec{w}_1(t) - \vec{w}_2(t)] \times \vec{r}(t) = 0 \tag{11}$$

valid for all the points of the body. Now, the standing assumption (made in the **Introduction** of this **Chapter**) that there are at least three non-collinear points of the body, we get that equation (11) is satisfied by three linearly independent vectors $\vec{r}(t)$. But by elementary vector algebra it follows then that $\vec{w}_1(t) - \vec{w}_2(t) = 0$.

This completes the proof of the uniqueness of the vector $\vec{w}(t)$.

Finally, we take $W(t)$ to be the line that goes along the vector $\vec{w}(t)$.

Comparison between the equations (4) and (5) makes it clear that the motion of the rigid body about the point of anchorage G is such that all the points of the body at the instant t are circulating about the instantaneous axis $W(t)$ with the common instantaneous angular speed $\|\vec{w}(t)\|$.

We have thus proved the following

Proposition 1 Suppose, a rigid body is free to rotate about a stationary point of it. Then at any instant t during the motion, there exists a unique line $W(t)$ passing through the fixed point and a unique vector $\vec{w}(t)$ in the direction of $W(t)$ which have the following property: Each point of the body appears to be rotating about the common instantaneous axis $W(t)$ with angular speed $\|\vec{w}(t)\|$.

Definition 1 The vector $\vec{w}(t)$ is the *instantaneous angular velocity* and the line $W(t)$, the *instantaneous axis of rotation* of the rigid body.

Next, we consider the unconstrained motion of the rigid body (unconstrained in the sense that no point of the body is unchored now.)

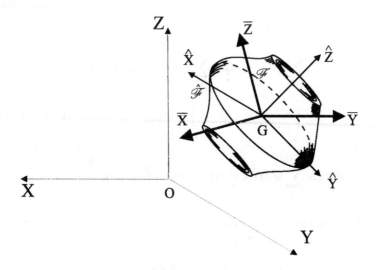

Figure 6.4

In addition to the frames $\mathcal{F}$ and $\hat{\mathcal{F}}$, we consider one more frame $\bar{\mathcal{F}}$ (to be abandoned soon). The frame $\bar{\mathcal{F}}$ has the following two properties: (i) origin of $\bar{\mathcal{F}} \equiv$ origin of $\hat{\mathcal{F}}$ but (ii) axes of $\bar{\mathcal{F}}$ are parallel to the respective axes of $\mathcal{F}$. Clearly $\bar{\mathcal{F}}$ moves along the body without undergoing rotational motion of the rigid body. Consequently we may consider the instantaneous angular velocity of the body to be the instantaneous angular velocity of the body frame $\hat{\mathcal{F}}$ relative to $\bar{\mathcal{F}}$ (about the common origin).

As usual $\vec{r}_G(t)$ is the instantaneous position vector of the point G of the body with respect to $\mathcal{F}$. Let P be any point of the body . Let $\vec{r}_p(t)$ be the position vector of P with respect to $\mathcal{F}$ and $\vec{s}_p(t)$ that with respect to $\hat{\mathcal{F}}$ (or equivalently with respect to $\bar{\mathcal{F}}$). Then we have: $\vec{r}_p(t) = \vec{r}_G(t) + \vec{s}_p(t)$. Differentiating this equation with respect to time, we get $\dot{\vec{r}}_p(t) = \dot{\vec{r}}_G(t) + \dot{\vec{s}}_p(t)$

Now note that the frame $\hat{\mathcal{F}}$ is rotating with respect to $\bar{\mathcal{F}}$ about the common origin and the point P is fixed in the rotating $\hat{\mathcal{F}}$ relative to $\bar{\mathcal{F}}$. Consequently the derivation (5) is applicable giving $\dot{\vec{s}}_p(t) = \vec{w}(t) \times \vec{s}_p(t)$. Substituting this result above, we get

$$\dot{\vec{r}}_p(t) = \dot{\vec{r}}_G(t) + \vec{w}(t) \times \vec{s}_p(t) \tag{12}$$

Instead of the points G and P we may consider any two points A and B of the body and then we may prove (exactly as the proof of (12)) the following result:

$$\dot{\vec{r}}_B(t) = \dot{\vec{r}}_A(t) + \vec{w}(t) \times \vec{r}_{AB}(t) \tag{13}$$

where $\vec{r}_A(t), \vec{r}_B(t)$ are the position vectors of A and B with respect to $\mathcal{F}$ and $\vec{r}_{AB}$ stands for $\vec{AB}$ at the instant t.

This result justifies our earlier remark that through the effects of rotational motion are felt differently at different reference points of the body, there is a single vector $\vec{w}(t)$ which determines the rotational state of the body.

Example 2 A, B and C are three non-collinear points of a rigid body. When they are at points $(a,0,0)$, $(0,\frac{a}{\sqrt{3}},0)(0,0,2a)$ at an instant, their respective velocities are $(u,0,0), (u,0,v)$ and $(u+v,-\sqrt{3}v,\frac{v}{2})$. Find the instantaneous angular velocity of the body.

Solution We have the equations

$$\dot{\vec{r}}_B(t) = \dot{\vec{r}}_A(t) + \vec{w}(t) \times \vec{r}_{AB}(t)$$
$$\dot{\vec{r}}_C(t) = \dot{\vec{r}}_A(t) + \vec{w}(t) \times \vec{r}_{AC}(t).$$

Substituting the given velocities in these equations we get

$$\left.\begin{array}{rcl} (u,0,v) & = & (u,0,0) + \vec{w}(t) \times (-a\frac{a}{\sqrt{3}},0) \\ (u+v,-\sqrt{3}\cdot v,\frac{v}{2}) & = & (u,0,0) + \vec{w}(t) \times (-a,0,2a) \end{array}\right\} \quad (*)$$

Let $\vec{w}(t)$ have the components (w_1, w_2, w_3) with respect to $\mathcal{F}$. Then we get $\vec{w}(t) \times (-a, \frac{a}{\sqrt{3}}, 0) = \left(-\frac{aw_3}{\sqrt{3}}, -aw_3, \frac{a}{\sqrt{3}}w_1 + aw_2\right)$ and $\vec{w}(t) \times (-a, 0, 2a) = (2aw_2, -aw_3 - 2aw_1, aw_2)$. Therefore, the equations $(*)$ take the following form

$$(u,0,v) = (u,0,0) + \left(\frac{-aw_3}{\sqrt{3}}, -aw_3, \frac{aw_3}{\sqrt{3}} + aw_2\right) \text{ and}$$

$$(u+v,-\sqrt{3}\cdot v,\frac{v}{2}) = (u,0,0) + (2aw_2, -aw_3 - 2aw_1, aw_2).$$

Rewriting them, we get

$$\left.\begin{array}{rcl} (-\frac{aw_3}{\sqrt{3}}, -aw_3, \frac{aw_1}{\sqrt{3}} & + & aw_2) = (0,0,v) \\ (2aw_2, -aw_3 -2aw_1, w_2a) & = & (v,-\sqrt{3}\cdot v,\frac{v}{2}). \end{array}\right\} \quad (**)$$

Separating the components, we get the following set of equation:
(i) $\frac{aw_3}{\sqrt{3}} = 0$, (ii) $aw_3 = 0$, (iii) $\frac{aw_1}{\sqrt{3}} + aw_2 = v$, (iv) $2aw_2 = v$, (v) $aw_3 - 2aw_1 = -\sqrt{3}v$, (vi) $w_2a = \frac{v}{2}$.

Equations (i) or (ii) give us $w_3 = 0$, (iv) gives $w_2 = \frac{v}{2a}$, (v) gives $w_1 = \frac{\sqrt{3}}{2a} \cdot v$. Hence $\vec{w}(t) = \left(\frac{\sqrt{3}}{2}\frac{v}{a}, \frac{v}{2a}, 0\right)$. Equivalently the angular speed of the body is $\frac{v}{a}$ and its direction is $\left(\frac{\sqrt{3}}{2}, \frac{1}{2}, 0\right)$. $\quad\square$

Example 3 A rigid body is free to rotate about a fixed point 0. Show that if $\vec{r}_A, \vec{r}_B$, are the position vectors of two points A and B of the body at an instant and if $\vec{v}_A$ and $\vec{v}_B$ are their velocity at the same instant, then the angular velocities $\vec{w}(t)$ of the body is given by

$$\vec{w}(t) = \frac{\vec{v}_A(t) \times \vec{v}_B(t)}{\vec{v}_A(t) \cdot \vec{r}_B(t)} = \frac{\vec{v}_B(t) \times \vec{v}_A(t)}{\vec{v}_B(t) \cdot \vec{r}_A(t)}$$

Solution By equation (5) we have

$$\vec{v}_A(t) = \vec{w}(t) \times \vec{r}_A(t) \text{and} \quad \vec{v}_B(t) = \vec{w}(t) \times \vec{r}_B(t).$$

Therefore,

$$\vec{v}_A(t) \times \vec{v}_B(t) = \vec{v}_A(t) \times (\vec{w}(t) \times \vec{r}_B(t)) = (\vec{v}_A(t) \cdot \vec{r}_B(t))\vec{w}(t) - (\vec{w}(t) \cdot \vec{v}_A(t))\vec{v}_B(t)$$
$$(*)$$

Now the result $\vec{v}_A(t) = \vec{w}(t) \times \vec{r}_A(t)$ implies that $\vec{v}_A(t)$ is perpendicular to $\vec{w}(t)$ and therefore $\vec{w}(t) \cdot \vec{v}_A(t) = 0$. Hence equation $(*)$ reduces to $\vec{v}_A(t) \times \vec{v}_B(t) = (\vec{v}_A(t) \cdot \vec{r}_B(t))\vec{w}(t)$ which in turn gives

$$\vec{w}(t) = \frac{\vec{v}_A(t) \times \vec{v}_B(t)}{\vec{v}_A(t) \cdot \vec{r}_B(t)}.$$

The other equation is proved similarly.   □

Example 4 Let n moving particles have position vectors $\vec{r}_1(t), \cdots, \vec{r}_n(t)$ at an instant t. Suppose further that there exists a single time dependent vector $\vec{\theta}(t)$ such that

$$\dot{\vec{r}}_j(t) = \vec{\theta}(t) \times \vec{r}_j(t) \qquad i \le j \le n.$$

Prove that the particles remain at constant relative distances during the motion.

Solution We consider the distances $\|\vec{r}_i(t) - \vec{r}_j(t)\|$. We have

$$\begin{aligned}
\frac{d}{dt}\left(\|\vec{r}_i(t) - \vec{r}_j(t)\|^2\right) &= \frac{d}{dt}\left(\vec{r}_i(t) - \vec{r}_j(t)\right) \cdot \left(\vec{r}_i(t) - \vec{r}_j(t)\right) \\
&= 2\left(\vec{r}_i(t) - \vec{r}_j(t)\right) \cdot \left(\dot{\vec{r}}_i(t) - \dot{\vec{r}}_j(t)\right) \\
&= 2\left(\vec{r}_i(t) - \vec{r}_j(t)\right) \cdot \left[\vec{\theta}(t) \times \left[\vec{r}_i(t) - \vec{r}_j(t)\right)\right] \quad (*)
\end{aligned}$$

[Since $\dot{\vec{r}}_i(T) = \vec{\theta}(t)\vec{r}_i(t), \dot{\vec{r}}_j(t) = \vec{\theta}(t) \times \vec{r}_j(t)$ implies $\left(\dot{\vec{r}}_i(t) - \dot{\vec{r}}_j(t)\right) = \theta(t)[\vec{r}_i(t) - \vec{r}_j(t)]$]

But $\vec{r}_i(t) - \vec{r}_j(t)$ is perpendicular to $\theta(\vec{t}) \times [\vec{r}_i(t) - \vec{r}_j(t)$ and so the scalar product on the right hand side of $(*)$ vanishes. Therefore the function $t \longmapsto \|\vec{r}_i(t) - \vec{r}_j(t)\|^2$ and hence the distance function $t \longmapsto \|r_i(t) - \vec{r}(t)\|$ is constant for any pair of the given particles. This proves the result.   □

Example 5 A small bead is threaded on a smooth wire in the shape of a curve given parametrically by $\vec{r} = \vec{r}(q)$ (q being the parameter). The wire rotates with constant angular velocity $\vec{w}$ about a vertical axis. Show

$$\frac{d}{dt}\left[\frac{1}{2}\dot{q}^2\left(\frac{d\vec{r}}{dq}\cdot\frac{d\vec{r}}{dq}\right)\right] = \dot{q}\frac{d\vec{r}}{dq}\cdot\ddot{r}$$

where $\dot{q}, \ddot{r}$ etc. are the time derivatives with respect to the frame rotating with the wire, $\vec{r}$ being taken with respect to the origin lying on the axis of rotation.

Deduce that $\left(\frac{d\vec{r}}{dq}\cdot\frac{d\vec{r}}{dq}\right)\cdot\dot{q}^2 - (\vec{w}\times\vec{r})\cdot(\vec{w}\times\vec{r}) - 2\vec{g}\cdot\vec{r} =$ constant . [This exercise is admittedly out of place in this chapter, because the problem is about a particle while the chapter is on the dynamics of rigid bodies. We have included the exercise here because it involves the newly introduced concept of the angular velocity.]

Solution We have

$$\frac{1}{2}\dot{q}^2\frac{d\vec{r}}{dq}\cdot\frac{d\vec{r}}{dq} = \frac{1}{2}\left(\dot{q}\frac{d\vec{r}}{dq}\right)\cdot\left(\dot{q}\frac{d\vec{r}}{dq}\right)$$

$$= \frac{1}{2}\left(\frac{d\vec{r}}{dt}\cdot\frac{d\vec{r}}{dt}\right) = \frac{1}{2}\dot{\vec{r}}\cdot\dot{\vec{r}}$$

Therefore,

$$\frac{d}{dt}\left(\frac{1}{2}\dot{q}^2\frac{d\vec{r}}{dq}\cdot\frac{d\vec{r}}{dq}\right) = \frac{d}{dt}\left(\frac{1}{2}\dot{\vec{r}}\cdot\dot{\vec{r}}\right)$$

$$= \dot{\vec{r}}\cdot\ddot{\vec{r}} = \dot{q}\left(\frac{d\vec{r}}{dq}\right)\cdot\ddot{r}$$

which proves the first equation.

The particle is acted upon by the gravity and therefore, its equation of motion is $m\ddot{\vec{r}} = m\vec{g}$. Now we have:

$$\frac{d}{dt}\left\{\dot{q}^2\frac{d\vec{r}}{dq}\cdot\frac{d\vec{r}}{dq} - (\vec{w}\times\vec{r})\cdot(\vec{w}\times\vec{r}) - 2\vec{g}\cdot\vec{r}\right\}$$

$$= 2\dot{\vec{r}}\cdot\ddot{\vec{r}} - 2(\vec{w}\times\vec{r})\cdot(\vec{w}\times\vec{r}) - 2\vec{g}\cdot\vec{r}$$

$$= 2\dot{\vec{r}}\cdot\vec{g} - 2(\vec{w}\times[\vec{w}\times\vec{r}])\cdot(\vec{w}\times\vec{r}) - 2\vec{g}\cdot\vec{r}$$

$$= 2\dot{\vec{r}}\cdot\vec{g} - 2(\vec{w}\times[\vec{w}\times\vec{r}])\cdot(\vec{w}\times\vec{r}) - 2\vec{g}\cdot\dot{\vec{r}}$$

$$= 0$$

(the middle term vanishes because $\vec{w}\times[\vec{w}\times\vec{r}]$ is orthogonal to $\vec{w}\times\vec{r}$).

This completes the proof of the second statement. $\square$

Example 6 A rigid body B has angular velocity $\vec{w}(t)$ and has one point held fixed at the origin of $\mathcal{F}$. If $\vec{w}(t)\times\dot{\vec{w}}(t)\neq 0$, for every t, show that the fixed point is the only point of the body having zero acceleration.

Solution Let P be any point of the body and let $\vec{r}(t)$ be its position vector. Note that $\vec{w}(t) \times \dot{\vec{w}}(t) \neq 0$ implies that $\left\{ \vec{w}(t), \dot{\vec{w}}(t), \vec{w}(t) \times \dot{\vec{w}}(t) \right\}$ is a linearly independent set of vectors and hence is a basis. We can therefore express $\vec{r}(t)$ as the following linear combination:

$$\vec{r}(t) = \alpha \vec{w}(t) + \beta \dot{\vec{w}}(t) + \gamma \vec{w}(t) \times \dot{\vec{w}}(t).$$

Now we have $\dot{\vec{r}}(t) = \vec{w}(t) \times \vec{r}(t)$ and therefore,

$$
\begin{aligned}
\ddot{\vec{r}} &= \dot{\vec{w}}(t) \times \vec{r}(t) + \vec{w}(t) \times \dot{\vec{r}}(t) \\
&= \dot{\vec{w}}(t) \times \vec{r}(t) + \vec{w}(t) \times [\vec{w}(t) \times \vec{r}(t)] \\
&= \dot{\vec{w}}(t) \times [\alpha \vec{w}(t) + \beta \dot{\vec{w}}(t) + \gamma \vec{w}(t) \times \dot{\vec{w}}(t)] \\
&\quad + \vec{w}(t) \times (\vec{w}(t) \times [\alpha \vec{w}(t) + \beta \dot{\vec{w}}(t) + \gamma \vec{w}(t) \times \dot{\vec{w}}(t)]) \\
&= \alpha \dot{\vec{w}}(t) \times \vec{w}(t) + \beta \dot{\vec{w}}(t) \times \dot{\vec{w}}(t) + \gamma \dot{\vec{w}} \times (\vec{w}(t) \times \dot{\vec{w}}(t)) \\
&\quad + \beta \vec{w}(t) \times (\vec{w}(t) \times \dot{\vec{w}}(t)) + \gamma \vec{w}(t) \times (\vec{w}(t) \times (\vec{w}(t) \times \dot{\vec{w}}(t)))\} \\
&= -\alpha \vec{w}(t) \times \dot{\vec{w}}(t) + 0 + \gamma \dot{w}(t)^2 \vec{w}(t) - \gamma(\vec{w}(t) \cdot \dot{\vec{w}}(t)) \dot{\vec{w}}(t) \\
&\quad + 0 + \beta(\vec{w}(t) \cdot \dot{\vec{w}}(t)) \vec{w}(t) - \beta w^2(t) \dot{\vec{w}}(t) \\
&\quad + \gamma \vec{w}(t) \cdot (\vec{w}(t) \times \dot{\vec{w}}(t)) \vec{w}(t) - \gamma w^2(t)(\vec{w}(t) \times \dot{\vec{w}}(t)). \\
&= -\alpha \vec{w}(t) \times \dot{\vec{w}}(t) + \gamma \dot{w}(t)^2 \vec{w}(t) - \gamma(\vec{w}(t) \cdot \dot{\vec{w}}(t)) \dot{\vec{w}}(t) \\
&\quad + \beta(\vec{w}(t) \cdot \dot{\vec{w}}(t)) \vec{w}(t) - \beta w(t)^2 \dot{\vec{w}}(t) - \gamma w^2(t)(\vec{w}(t) \times \dot{\vec{w}}(t)) \\
&= [\gamma \dot{w}(t)^2 + \beta \vec{w}(t) \cdot \dot{\vec{w}}(t)] \vec{w}(t) \\
&\quad + [-\gamma(\vec{w}(t) \cdot \dot{\vec{w}}(t)) - \beta w^2(t)] \dot{\vec{w}}(t) \\
&\quad + [-\gamma w^2(t) - \alpha] \vec{w}(t) \times \dot{\vec{w}}(t).
\end{aligned}
$$

Thus, we have derived:

$$
\begin{aligned}
\ddot{\vec{r}}(t) &= \left[\gamma \dot{w}(t)^2 + \beta(\vec{w}(t) \cdot \dot{\vec{w}}(t)) \right] \vec{w}(t) \\
&\quad - \left[\gamma(\vec{w}(t) \cdot \dot{\vec{w}}(t)) + \beta w(t)^2 \right] \dot{\vec{w}}(t) \\
&\quad - \left[\gamma w^2(t) + \alpha \right] \vec{w}(t) \times \dot{\vec{w}}(t).
\end{aligned}
$$

Consequently if the point P has zero acceleration, then the above equation gives rise to the following three simultaneous equations in $\alpha, \beta \gamma$:

$$\gamma \dot{w}(t)^2 + \beta(\vec{w}(t) \cdot \dot{\vec{w}}(t)) = 0 \qquad (*).$$

$$\gamma(\vec{w}(t) \cdot \dot{\vec{w}}(t)) + \beta w(t)^2 = 0 \qquad (**)$$

$$\alpha + \gamma w(t)^2 = 0 \qquad (***)$$

Solving $(*)$ and $(**)$ for γ, we get $\gamma \left[w(t)^2 \dot{w}(t)^2 - (\vec{w}(t) \cdot \dot{\vec{w}}(t)^2 \right] = 0$. Note that $\vec{w}(t) \times \dot{\vec{w}}(t) \neq 0$ implies $w^2(t) \dot{w}(t)^2 \neq (\vec{w}(t) \cdot \dot{\vec{w}}(t))^2$ and consequently, the equation $\gamma \left[w(t)^2 \dot{w}(t)^2 - (\vec{w}(t) \cdot \dot{\vec{w}}(t))^2 \right] = 0$ implies $\gamma = 0$.

Substituting $\gamma = 0$ in (**) gives $\beta = 0$. Also, substitution of $\gamma = 0$ in (***) gives $\alpha = 0$. Thus, $\alpha = \beta = \gamma = 0$ and so, the position vector having zero acceleration must be the fixed point zero. $\qquad\square$

6.4 Equations of Motion

It was noted earlier that the general motion of a rigid body is a combined effect of the following two motions:

(1) The motion of the center of mass of the rigid body.

(2) The rotational motion of the body about its center of mass.

The motion(1) is governed by the equation:

$$M\frac{d\vec{V}}{dt}(t) = F^{(e)} \qquad (14)$$

where M is the total mass of the body, considered to be concentrated at the center of mass; $\vec{V}(t)$, being the velocity of the center of mass and $F^{(e)}$ being the total external force, considered to be acting on the body at the center of mass. To deal with the rotational motion of the body, we have designated the angular momentum $\vec{L} = \vec{L}_G$. We have also considered the torque $\vec{N} = \vec{N}_G$ of the forces acting on the body. Now we have the equation of rotational motion.

$$\frac{d\vec{L}}{dt}(t) = \vec{N}(t). \qquad (15)$$

Equations (14) and (15) are the equations of motion of the rigid body.

Since(as noted in **Chapter 5**) equation (14) can be interpreted as the equation of motion of a particle and as such we have some ideas regarding how to solve it. In fact, we have been studying some techniques about solving the equation of motion of a single particle throughout **Chapters 1 to 4** . We will therefore consider the problem of motion of G to be manageable and so we will concentrate on the rotational motion only. This we do by making a simplifying assumption that the body is anchored at a stationary point of the body and thus we will study the rotational motion of it around the point of anchorage.

At this stage, we compare the equations (14) and (15). on the right hand sides, the torque $\vec{N}$ is the counterpart of the force F. On the left hand sides the angular momentum $\vec{L}$ corresponds to the linear momentum $\vec{M}$ and the instantaneous angular velocity $\vec{w}(t)$ corresponds to the (linear) velocity $\vec{V}$. We therefore expect a suitable coefficient of the angular velocity $\vec{w}$ (perhaps an operator, rather than a scalar factor) that will render the angular momentum to $\vec{L}$ in a product form just as the linear momentum is in the form $M \cdot \vec{V}$. If such a representation (that is $\vec{L} = (*) \cdot \vec{w}$ is possible, then equation (15) will bear similarity with equation (14).

We will see below that this can indeed be accomplished by transferring equation (15) to the body frame of reference $\mathcal{F}$.

Let us begin by resolving the angular velocity $\vec{w}(t)$ along the axes of $\hat{\mathcal{F}}$:

$$\vec{w}(t) = \Omega_1(t)\vec{u}_1 + \Omega_2(t)\vec{u}_2 + \Omega_3(t)\vec{u}_3. \tag{16}$$

6.5   The Inertia Tensor

We are considering the rotational motion of a rigid body about a stationary point. As in the previous sections, $\mathcal{F}$ is a stationary frame, $\hat{\mathcal{F}}$ is a body frame both having their origins at the fixed point.

First, we consider the discrete model (2) of the rigid body.

Let $\vec{r}_j$ be the position vector of the particle P_j with respect to $\mathcal{F}$ and let $\vec{s}_j(t)$ that with respect to $\hat{\mathcal{F}}$. We write $\vec{s}_j = x_j^1 \vec{u}_1 + x_j^2 \vec{u}_2 + x_j^3 \vec{u}^3$ so that (x_j^1, x_j^2, x_j^3) are the coordinates of P in the frame $\hat{\mathcal{F}}$

To begin with, we consider the angular momentum $\vec{L} = \vec{L}_o$ of the body about the fixed point0. We have:

$$
\begin{aligned}
\vec{L}(t) &= \sum_{\ell=1}^{n} m_\ell \vec{r}_\ell(t) \times \dot{\vec{r}}_\ell(t) \\
&= \sum_{\ell=1}^{n} m_\ell \vec{r}_\ell(t) \times [\vec{w}(t) \times \vec{r}_\ell(t)] \\
&= \sum_{\ell=1}^{n} m_\ell r_\ell^2(t)\vec{w}(t) - (\vec{w}(t) \cdot \vec{r}_\ell(t))\vec{r}_\ell(t)
\end{aligned}
$$

Now, we have $r_\ell^2 = s_\ell^2 = \sum_{k=1}^{3}(x_\ell^k)^2$. Similarly, $\vec{w}(t) \cdot \vec{r}_\ell(t) = \sum_{i=1}^{3} \Omega_i(t)x_\ell^i$. Therefore, $\vec{L}(t)$ is given by

$$\vec{L}(t) = \sum_{\ell=1}^{n} m_\ell \left[\sum_{i=1}^{3}(x_\ell^i)^2 \cdot \sum_{k=1}^{3} \Omega_k(t)\vec{u}_k - \sum_{k=1}^{3} \Omega_k(t)x_\ell^k \sum_{j=1}^{3} x_\ell^j \vec{u}_j \right].$$

We now consider the components of the $\vec{L}(t)$ along the axes of $\hat{\mathcal{F}} : \vec{L}(t) = \sum_{j=1}^{3} L_j(t)\vec{u}_j$. Then in the above equation the jth component of $\vec{L}(t)$ along $\vec{u}_j$ is given by

$$L_j(t) = \sum_{\ell=1}^{n} m_\ell \left\{ \sum_{i=1}^{3}(x_\ell^i)^2 \Omega_j(t) - \left(\sum_{k=1}^{3} \Omega_k(t)x_\ell^k \right) x_\ell^j \right\}.$$

Substituting $\Omega_j(t) = \sum_{k=1}^{3} \delta_{jk}\Omega_k(t)$ in the above equation, we get

$$
\begin{aligned}
L_j(t) &= \sum_{\ell=1}^{n} m_\ell \left\{ \sum_{i=1}^{3} (x_\ell^i)^2 \cdot \sum_{k=1}^{3} \delta_{jk}\Omega_k(t) - \sum_{k=1}^{3} \Omega_k(t) x_\ell^k x_\ell^j \right\}. \\
&= \sum_{\ell=1}^{n} m_\ell \left\{ \sum_{k=1}^{3} \left[\delta_{jk} \sum_{i=1}^{3} (x_\ell^i)^2 - x_\ell^k x_\ell^j \right] \Omega_k(t) \right\} \\
&= \sum_{k=1}^{3} \left\{ \sum_{\ell=1}^{n} m_\ell \left[\delta_{jk} \sum_{i=1}^{3} (x_\ell^i)^2 - x_\ell^j x_\ell^k \right] \right\} \Omega_k(t).
\end{aligned}
$$

We put

$$
I_{jk} = \sum_{\ell=1}^{n} m_\ell \left\{ \delta_{jk} \sum_{i=1}^{3} (x_\ell^i)^2 - x_\ell^j x_\ell^k \right\}. \tag{17}
$$

Now we have obtained above the equation: $L_j(t) = \sum_k I_{jk}\Omega_k(t)$ $1 \le j \le 3$. Combining these equations in a matrix form we get:

$$
\begin{bmatrix} L_1(t) \\ L_2(t) \\ L_3(t) \end{bmatrix} = \begin{bmatrix} I_{11}, & I_{12}, & I_{13} \\ I_{21}, & I_{22}, & I_{23} \\ I_{31}, & I_{32}, & I_{33} \end{bmatrix} \begin{bmatrix} \Omega_1(t) \\ \Omega_2(t) \\ \Omega_3(t) \end{bmatrix}
$$

or in short notation

$$
\vec{L}(t) = I(\vec{w}(t)) \tag{18}
$$

where I is the 3×3 matrix $[I_{ij}]$ $1 \le i, j \le 3$ described above.

Note that equation (18) refer the vector $\vec{L}(t)$ (having components $(L_1(t), L_2(t), L_3(t))$ to the body frame $\hat{\mathcal{F}}$. Similarly $\vec{\Omega}(t) = (\Omega_1(t)\vec{u}_1 + \Omega_2(t)\vec{u}_2 + \Omega_3(t)\vec{u}_3$ is the representation of the instantaneous angular velocity with respect to $\hat{\mathcal{F}}$.

We were considering the discrete model of the body. Obviously, in case of the continuous model (B, ρ) the outer summation in(17) which extends over all the particles of the body should be replaced by an integration over the rigid body B. In the process, the discrete masses $m_1, m_2, \cdots, m_n$ of the particle are to be replaced by the continuous mass density factor $P \longmapsto \rho(P)$. Thus, we replace the particles P_j having mass m_j and position vector $\vec{s}_j$ by a small volume element $dv(P)$ around a point $P(\simeq \vec{s}(P))$ having mass $\rho(P) \cdot dv(P)$, and so on. This consideration leads us to the formula

$$
I_{jk} = \int_B \rho(P) \left[\delta_{jk} \sum_{i=1}^{3} (x^i(P))^2 - x^i(P) \cdot x^j(P) \right] dv(P) \tag{19}
$$

The matrix $I = [I_{jk}]_{i \le j, k \le 3}$ plays an important role in the dynamics of the body. It turns out that this 3×3 matrix is the rotational counter-part of the total mass of the body in the comparison between the rotational and translational motions. Note that the matrix depends on the choice of the body frame $(I = I(\hat{\mathcal{F}}))$.

Definition 2 $I = [I_{jk}]$ is called the **inertia matrix** or the **inertia tensor** of the rigid body relative to the body frame $\hat{\mathcal{F}}$. The diagonal terms I_{11}, I_{22}, I_{33} are called the **moments of inertia** of the body about the X, Y and Z axis of $\hat{\mathcal{F}}$. The remaining, off-diagonal terms of the matrix are called the **products of inertia**.

We now use the inertia tensor and obtain an expression for the kinetic energy of the rotating body. We consider the discrete model first. We have

$$
\begin{aligned}
T(t) &= \frac{1}{2} \sum_{\ell=1}^{n} m_\ell \|\dot{\vec{r}}_\ell(t)\|^2 \\
&= \frac{1}{2} \sum_{\ell=1}^{n} m_\ell \|\vec{w}(t) \times \vec{r}_\ell\|^2 \\
&= \frac{1}{2} \sum_{\ell=1}^{n} m_\ell \left[w^2(t) \cdot r_\ell^2 - (\vec{w}(t) \cdot \vec{r}_\ell(t))^2 \right] \\
&= \frac{1}{2} \sum_{e=1}^{n} m_\ell \left[\sum_{j=1}^{3} \Omega_j^2 \cdot \sum_{i=1}^{3} (x_\ell^i)^2 - (\sum_{j,k=1}^{3} \Omega_j \Omega_k x_\ell^j x_\ell^k \right] \\
&= \frac{1}{2} \sum_{\ell=1}^{n} m_\ell \left[\sum_{jk=1}^{3} \delta_{jk} \cdot \Omega_j \Omega_k \sum_{i=1}^{3} (x_\ell^i)^2 - \sum_{j,k} \Omega_j \Omega_k x_\ell^j x_\ell^k \right] \\
&= \frac{1}{2} \sum_{\ell=1}^{n} m_\ell \left[\sum_{jk=1} \left\{ \delta_{jk} \sum_{i=1}^{3} (x_\ell^i)^2 - x_\ell^j x_\ell^k \right\} \Omega_j \Omega_k \right] \\
&= \frac{1}{2} \sum_{jk=1}^{3} \left\{ \sum_{\ell} m_\ell \left[\delta_{jk} \sum_{i=1}^{3} (x_\ell^i)^2 - x_\ell^j x_\ell^k \right] \right\} \Omega_j \Omega_k \\
&= \frac{1}{2} \sum_{jk=1}^{3} I_j k \Omega_j \Omega_k.
\end{aligned}
$$

Thus, we have proved the formula,

$$
T(t) = \frac{1}{2} \sum_{j,k=1}^{3} I_{jk} \Omega_j(t) \Omega_k(t) \tag{20}
$$

Same formula can be derived easily in case of the continuous model. All that need be done is to replace the discrete summation (over the particles) by an integration extending over the rigid body. We leave the details for the reader to work out as an exercise.

An important property of the inertia tensor is that it is symmetric $I_{ij} = I_{ji}$; $1 \leq i, j \leq n$. This is evident from formulae (19) and (20). Now, according to a result of linear algebra, we can diagonalize the inertia matrix (because it is symmetric) by rotating the body frame without changing the origin. In other words, we have the result:

Given any point 0 of the rigid body, we can always choose a body frame $\hat{\mathcal{F}}$ with its origin at 0 in such a way that the off-diagonal terms of the associated interia metrix are all zero:

$$I = \begin{bmatrix} I_{11} & 0 & 0 \\ 0 & I_{22} & 0 \\ 0 & 0 & I_{33} \end{bmatrix}$$

We will call the directions $\vec{u}_1, \vec{u}_2, \vec{u}_3$ of the axes of such a body frame, a triple of **principle directions** through 0. Also, we call the associated moments of inertia (I_{11}, I_{22}, I_{33}) the **principal moments of inertia**. We will denote them (for the sake of simplicity) by I_1, I_2, I_3 instead of I_{11}, I_{22}, I_{33}.

Clearly, a body frame with principal directions as its axes simplifies many of the calculations. For example, we have the following simplified expressions for angular momentum and kinetic energy in such a frame.

(a) $\vec{L} = I_1 \Omega_1(t) \vec{u}_1 + I_2 \Omega_2(t) \vec{u}_2 + I_3 \Omega_3(t) \vec{u}_3$.

(b) $T(t) = \frac{1}{2} \left[I_1 \Omega_1^2(t) + I_2 \Omega_2^2(t) + I_3 \Omega_3(t)^2 \right]$

Because such frames of reference (in the body) are always available at any point 0, in the following, we will make use of only such frames.

In an elementary course on classical mechanics, the reader had probably come across a simpler situation in which the body rotates about a stationary axis. Suppose, $\vec{e}$ is a unit vector along this axis. Then a quantity $I(\vec{e})$ called moment of inertia about the line $\vec{e}$ standing for the following expression was considered:

(1) **Discrete case:** $I(\vec{e}) = \sum_{\ell=1}^{n} m_\ell (d_\ell)^2$ where d_ℓ is the distance of the particle P_ℓ from the axis of rotation that is the line having direction $\vec{e}$.

(2) **Continuous case:** $I(\vec{e}) = \int_B \rho(P) d^2(P) d\nu(P)$ where $d(P)$ stands for the distance of a point P of the body from the axis of rotation.

Note that in terms of the present notation, the principal moments of inertia I_1, I_2, I_3 are $I(\vec{u}_1), I(\vec{u}_2)$ and $I(\vec{u}_3)$ respectively for any body frame $\hat{\mathcal{F}}$ (with principal axes). This explains the nomenclature *principal moments of inertia*. Also, recall that the kinetic energy of the body in the above envisaged situation is given by

$$T(t) = \frac{1}{2} I(\vec{e}) \Omega^2(t)$$

where $\vec{\Omega}(t) = \Omega(t)\vec{e}$.

Example 7 Let (ℓ_x, ℓ_y, ℓ_z) be the direction cosines of a unit vector $\vec{e}$ with respect to a body frame $\hat{\mathcal{F}}$) (that is, ℓ_x is the cosine of angle between $\vec{e}$ and X-axis of $\mathcal{F}$ and so on). Prove:

$$I(\vec{e}) = I_1 \ell_x^2 + I_2 \ell_y^2 + I_3 \ell_z^2.$$

Solution We consider the angular velocity $\vec{\Omega} = \Omega\vec{e}$ with $(\Omega \neq 0)$. We have $\vec{e} = \ell_x \vec{u}_1 + \ell_y \vec{u}_2 + \ell_z \vec{u}_3$ and therefore $\vec{\Omega} = \Omega\ell_x \vec{u}_1 + \Omega\ell_y \vec{u}_2 + \Omega\ell_z \vec{u}_3$. Now we consider the associated kinetic energy:

$$
\begin{aligned}
T(t) &= \frac{1}{2}\left[I_1 \ell_x^2 \Omega^2(t) + I_2 \ell_y^2 \Omega^2(t) + I_3 \ell_z^2 \Omega_3^2(t)\right] \\
&= \frac{1}{2}\left[I_1 \ell_x^2 + I_2 \ell_y^2 + I_3 \ell_z^2\right] \Omega^2(t). \qquad (*)
\end{aligned}
$$

Also, we have

$$T(t) = \frac{1}{2}I(e)\Omega^2(t). \qquad (**)$$

Now, comparison between the two results $(*)$ and $(**)$ give the required result. $\square$

The above example makes it clear that the quantity $I(\vec{e})$ is sufficient (instead of the whole matrix $[I_{ij}]$) when the body is rotating in a particular direction $\vec{e}$ while the inertia matrix is a dynamical operator which is capable of considering rotations along arbitrary directions.

Moreover, given a point 0 of the body, to compute the inertia tensor, all we have to do is the following:

(i) Taking into consideration the geometry of the rigid body and the mass distribution in it, make a suitable choice of a set $\vec{u}_1, \vec{u}_2, \vec{u}_3$ of principal directions.

(ii) Calculate the moments of inertia $I(\vec{u}_1), I(\vec{u}_2), I(\vec{u}_3)$.

(iii) Finally, put them along the diagonal of the 3×3 matrix in their natural order, the remaining (i.e. the off-diagonal) terms being all equal to zero.

Alternatively, we may calculate the inertia tensor with respect to any body frame of reference having its origin at the chosen point 0 of the body. This 3×3 matrix will provide (i) three (distinct or equal) eigenvalues, (call them say J_1, J_2, J_3) and (ii) three mutually perpendicular unit vectors (call them $\vec{v}_1, \vec{v}_2, \vec{v}_3$) namely the eigenvectors of the matrix. Then the frame $\vec{\mathcal{F}}$ having origin at 0, axes along $\vec{v}_1, \vec{v}_2, \vec{v}_3$ is a body frame with principal directions such that the inertia matrix of the body with respect to $\vec{\mathcal{F}}$ is

$$
\begin{bmatrix}
J_1 & 0 & 0 \\
0 & J_2 & 0 \\
0 & 0 & J_3
\end{bmatrix}.
$$

6.6   Euler's Equations

The analysis of motion of a rigid body set forth so far gives rise to the following comparison between the translational and rotational features of the motions:

Translational Motion	Rotational Motion
The velocity $\vec{V}(t)$ of the center of mass of the body	The angular velocity $\vec{w}(t)$ of the body
The linear momentum $\vec{p}(t) = M\vec{V}(t)$	The angular momentum $\vec{L} = I\vec{w}$
The total mass M of the body	The inertia tensor I of the body
The total external force acting on the body	The torque $\vec{N}$ of the external forces acting on the body
Equation of translational motion of the body $M\frac{d\vec{V}}{dt} = F.$	Equation of rotational motion of the body. $\frac{d\vec{L}}{dt} = N.$

This comparison suggests that the equation $\frac{d\vec{L}}{dt} = \vec{N}$ could be rewritten in a way similar to its translational counterpart by bringing in the inertia tensor I. However, the inertia tensor is defined relative to the body frame $\hat{\mathcal{F}}$ while the equation $\frac{d\vec{L}}{dt} = \vec{N}$ is in the frame $\mathcal{F}$. Therefore, the equation must be transferred to $\hat{\mathcal{F}}$.

Now in $\mathcal{F}$ we have the following:

$$\left.\begin{array}{ll} (a) & \vec{w}(t) = \Omega_1(t)\vec{u}_1(t) + \Omega_2(t)\vec{u}_2(t) + \Omega_3(t)\vec{u}_3(t) \\ (b) & \dot{\vec{u}}_i(t) = \vec{w}(t) \times \vec{u}_i(t), \quad i = 1, 2, 3 \\ (c) & \vec{L}(t) = I_1\Omega_1(t)\vec{u}_1(t) + I_2\Omega_2(t)\vec{u}_2(t) + I_3\Omega_3(t)\vec{u}_3(t). \end{array}\right\} (21)$$

Combining (a) and (b) we get:

$$\begin{aligned} \dot{\vec{u}}_1(t) &= [\Omega_1(t)\vec{u}_1(t) + \Omega_2(t)\vec{u}_2(t) + \Omega_3(t)\vec{u}_3(t)] \times \vec{u}_1(t) \\ &= -\Omega_3(t)\vec{u}_2(t) + \Omega_2(t)\vec{u}_3(t). \end{aligned}$$

Thus we have:

(i) $\dot{\vec{u}}_1(t) = \Omega_3(t)\vec{u}_2(t) - \Omega_2(t)\vec{u}_3(t).$

Similarly we have

(ii) $\dot{\vec{u}}_2(t) = -\Omega_3(t)\vec{u}_1(t) + \Omega_1(t)\vec{u}_3(t)$

(iii) $\dot{\vec{u}}_3(t) = \Omega_2(t)\vec{u}_1(t) - \Omega_1(t)\vec{u}_2(t)$

Next, we differentiate equation (c) above to get

$$
\begin{aligned}
\dot{\vec{L}}(t)(t) &= I_1\dot{\Omega}_1(t)\vec{u}_1(t) + I_1\Omega_1(t)\dot{\vec{u}}_1(t) + I_2\dot{\Omega}_2(t)\vec{u}_2(t)) + I_2\Omega_2(t)\dot{\vec{u}}_2(t) \\
&\quad I_3\dot{\Omega}_3(t)\vec{u}_3(t) + I_3\Omega_3(t)\dot{\vec{u}}_3(t) \\
&= I_1\dot{\Omega}_1(t)\vec{u}_1(t) + I_2\dot{\Omega}_2(t)\vec{u}_2(t) + I_3\dot{\Omega}_3(t)\vec{u}_3(t) \\
&\quad + I_1\Omega_1(t)\left[\Omega_3(t)\vec{u}_2(t) - \Omega_2(t)\vec{u}_3(t)\right] \\
&\quad + I_2\Omega_2(t)\left[\Omega_3(t)\vec{u}_1(t) - \Omega_1(t)\vec{u}_3(t)\right] \\
&\quad + I_3\Omega_3(t)\left[\Omega_2(t)\vec{u}_1(t) - \Omega_1(t)\vec{u}_2(t)\right]
\end{aligned}
$$

the last three lines of the above equation being obtained by substituting (21)in place of $\dot{\vec{u}}_1(t), \dot{\vec{u}}_2(t), \dot{\vec{u}}_3(t)$. Regrouping terms, we get

$$
\begin{aligned}
\frac{d\vec{L}}{dt} &= \left[I_1\dot{\Omega}_1(t) + (I_3 - I_2)\Omega_2(t)\Omega_3(t)\right]\vec{u}_1(t) \\
&\quad + \left[I_2\dot{\Omega}_2(t) + (I_1 - I_3)\Omega_1(t)\Omega_3(t)\right]\vec{u}_2(t) \\
&\quad + \left[I_3\dot{\Omega}_3(t) + (I_2 - I_1)\Omega_1(t)\Omega_2(t)\right]\vec{u}_3(t)
\end{aligned}
$$

We resolve the torque $\vec{N}(t)$ along the axes of $\hat{\mathcal{F}}$

$$
\vec{N}(t) = N_1(t)\vec{u}_1(t) + N_2(t)\vec{u}_2(t) + N_3(t)\vec{u}_3(t).
$$

Now the equation $\frac{d\vec{L}}{dt}(t) = \vec{N}(t)$ takes the form

$$
\begin{aligned}
&\left[I_1\dot{\Omega}_1(t) + (I_3 - I_2)\Omega_2(t)\Omega_3(t)\right]\vec{u}_1(t) \\
&+ \left[I_2\dot{\Omega}_2(t) + (I_1 - I_3)\Omega_1(t)\Omega_3(t)\right]\vec{u}_2(t) \\
&+ \left[I_3\dot{\Omega}_3(t) + (I_2 - I_1)\Omega_1(t)\Omega_2(t)\right]\vec{u}_3(t) \\
&= N_1(t)\vec{u}_1(t) + N_2(t)\vec{u}_2(t) + N_3(t)\vec{u}_3(t)
\end{aligned}
$$

Separating the components of the above equations, we get

$$
\left.
\begin{aligned}
I_1\dot{\Omega}_1(t) + (I_3 - I_2)\Omega_2(t)\Omega_3(t) &= N_1(t) \\
I_2\dot{\Omega}_2(t) + (I_1 - I_3)\Omega_1(t)\Omega_3(t) &= N_2(t) \\
I_3\dot{\Omega}_3(t) + (I_2 - I_1)\Omega_1(t)\Omega_2(t) &= N_3(t)
\end{aligned}
\right\}
\qquad (22)
$$

Equations (22) are called "Euler's equations" of rotational motion of a rigid body. Note that the equations are differential equations in the components $\Omega_1(t), \Omega_2(t), \Omega_3(t)$ of the angular velocity of the body in the body frame $\hat{\mathcal{F}}$ (and not in the stationary frame $\mathcal{F}$). In other words we have transferred the equation $\frac{d\vec{L}}{dt}(t) = \vec{N}$ (which is valid in the stationary frame $\mathcal{F}$) to the body frame $\hat{\mathcal{F}}$.

The idea of getting the differential equation of motion in the body frame $\hat{\mathcal{F}}$ (rather than in the stationary frame $\mathcal{F}$) is due to Leonard Euler and hence the equations are called Euler's equations.

Example 8 A circular cylinder of mass M, radius a is rolling down an inclined plane. The cylinder is rolling without slipping, its axis remaining horizontal. Let J be the moment of inertia of the cylinder about its axis and let α be the angle of elevation of the inclined plane. Find the acceleration of the cylinder.

Solution Let $x(t)$ be the distance covered and $\theta(t)$ the angle turned by the cylinder about its axis in time t. Then we have $x(t) = a\theta(t)$. The instantaneous angular velocity of the byliner is $\dot{\theta}(t)\vec{j}$ where; $\vec{j}$ is a (constant) unit vector along the axis of the cylinder. Consequently, $\vec{L}(t) = J\dot{\theta}(t)\vec{j}$.

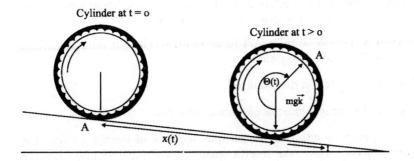

Figure 6.5

Note that forces acting on the cylinder are (a) the force of friction F and (b) the component $M \cdot g \cdot \sin\alpha$ of the weight of the cylinder along the inclined plane. Therefore, the equation of rotational motion of the cylinder is :

$$M\ddot{x}(t) = -F + Mg\sin\alpha \qquad (*)$$

The force of friction, F gives rise to the torque $\vec{N} = aF\vec{j}$. Now the equation of rotational motion is

$$\frac{d}{dt}(J\dot{\theta}\vec{j}) = aF\vec{j} \qquad (**)$$

which simplifies to

$$J\ddot{\theta} = a \cdot F. \qquad (**)$$

But $x = a\theta$ and so $\ddot{x} = a\ddot{\theta}$. that is $\ddot{\theta} = \frac{\ddot{x}}{a}$. Consequently, equation $(**)$ becomes

$$\frac{J\ddot{x}}{a^2} = F \qquad (***)$$

Combining $(*)$ and $(***)$ we get

$$M\ddot{x} + J\frac{\ddot{x}}{a^2} = Mg\sin\alpha$$

and hence $\ddot{x} = \frac{g\sin\alpha}{(1 + \frac{J}{Ma^2})}$ which is the required acceleration.

Note that the larger the ratio $\frac{J}{Ma^2}$, the slower is the motion. In particular, the motion depends on the radius of the cylinder.   □

6.7   Symmetric Rigid Bodies

It is very difficult to integrate equations (22) for a body having arbitrary shape and arbitrary mass distribution. But if the mass distribution is uniform and if the shape of the body has geometric symmetries, then the equations are considerably simplified. We consider a particular case of such a situation described in the definition given below:

Definition 3 A rigid body is said to be **symmetric** if it admits a frame of reference $\hat{\mathcal{F}}$ in such a way that

(i) The inertia metric is diagonalized.

(ii) Two of the three moments of inertia are equal.

The axis corresponding to the unequal moment of inertia is called the **symmetry axis**. Thus, the inertia tensor of a symmetric body has the form

$$\begin{bmatrix} A & 0 & 0 \\ 0 & A & 0 \\ 0 & 0 & B \end{bmatrix}$$

The body Z-axis in this case is the symmetry axis of the body. Here are some examples of symmetric bodies.

(I) A sphere with uniform mass distribution, total mass M, raidus a with respect to any body frame having its origin at the center of the sphere. The inertia matrix of the sphere is given by:

$$I = \begin{bmatrix} \frac{2}{5}Ma^2 & 0 & 0 \\ 0 & \frac{2}{5}Ma^2 & 0 \\ 0 & 0 & \frac{2}{5}Ma^2 \end{bmatrix}$$

(II) A cube having uniform mass distribution; total mass M, edge length a, has the inertia matrix

$$I = \begin{bmatrix} \frac{2}{3}Ma^2 & 0 & 0 \\ 0 & \frac{2}{3}Ma^2 & 0 \\ 0 & 0 & \frac{2}{3}Ma^2 \end{bmatrix}$$

with respect to any body frame $\hat{\mathcal{F}}$ having its origin at the center and axes parallel to the edges of the cube. Any straight line through the center and parallel to an edge is a symmetry axis.

(III) A circular cylinder of radius a, height h total mass M, with uniform mass distribution has the inertia matrix

$$\begin{bmatrix} \frac{m}{2}a^2 & 0 & 0 \\ 0 & \frac{m}{12}(h^2 + 3a^2) & 0 \\ 0 & 0 & \frac{m}{12}(h^2 + 3a^2) \end{bmatrix}$$

with respect to any body frame $\hat{\mathcal{F}}$ having its origin at the center of the cylinder and the z-axis along axis of the cylinder. Clearly, the axis of the cylinder is the symmetric axis.

(IV) A spinning top is obviously a symmetric body.

We consider a symmetric rigid body with its inertia tensor (with respect to a body frame $\hat{\mathcal{F}}$).

$$I = \begin{bmatrix} A & 0 & 0 \\ 0 & A & 0 \\ 0 & 0 & B \end{bmatrix}$$

with $A \neq B$. Suppose the body rotates freely about the origin 0 of $\hat{\mathcal{F}}$. Also suppose, there are no forces acting on the body. We describe below its rotational motion.

Now we have $\vec{N}(t) \equiv 0$ that is $N_1(t) = N_2(t) = N_3(t) \equiv 0$. Euler's equations take the following form:

(a) $A\dot{\Omega}_1(t) + (B - A)\Omega_2(t)\Omega_3(t) = 0$.

(b) $A\dot{\Omega}_2(t) + (A - B)\Omega_1(t)\Omega_3(t) = 0$.

(c) $B\dot{\Omega}_3(t) + (A - A)\Omega_1(t)\Omega_2(t) = 0$.

Equation (c) gives $\dot{\Omega}_3(t) = 0$ and hence $\Omega_3(t)$ is a constant. We denote its constant value by Ω_3. Assume that $B > A$. (The other case $B < A$ is left as the Exercise 11 at the end of this chapter.) Denoting the constant $\left(\frac{B-A}{A}\right)\Omega_3$ by η, the above equations (a) and (b) become:
(a) $\dot{\Omega}_1(t) = \eta\,\Omega_2(t)$ and (b) $\dot{\Omega}_2(t) = -\eta\,\Omega_1(t)$. Clearly, this pair of equations has the solutions:

$$\Omega_1(t) = \alpha \cdot \sin(\sqrt{\eta} \cdot t + \beta), \Omega_2(t) = \alpha \cos(\sqrt{\eta} \cdot t + \beta) \qquad (*)$$

where α and β are some constants of integration. Hence the angular velocity $\vec{w}(t)$ is given (in terms of its components along the axes of the body frame $\hat{\mathcal{F}}$) by

$$\vec{w}(t) = \alpha \, \sin(\sqrt{\eta} \cdot t) + \beta)\vec{u}_1(t) + \alpha \cos(\sqrt{\eta} \cdot t + \beta)\vec{u}_2(t) + \Omega_3\vec{u}_3(t).$$

The magnitude of the angular velocity $w(t)$ is $\sqrt{\alpha^2 + \Omega_3^2}$ and hence constant. Consequently (i.e. because its z-component is constant while its X- and Y-components are given in ($*$)), we see that the angular velocity vector $\vec{w}(t)$ as seen from the body frame $\hat{\mathcal{F}}$, describes a cone about the symmetry axis. It precesses about the symmetry axis with constant angular velocity η in a counter clockwise sense.

In the stationary frame $\mathcal{F}$ we have two constants of motion:

1. The angular momentum $\vec{L}$.

2. the total energy E which is the kinetic energy $\frac{1}{2}\vec{L} \cdot \vec{w}(t)$.

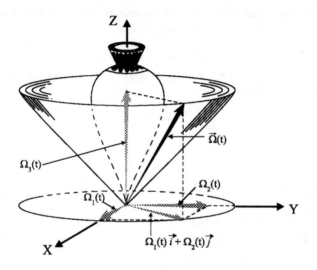

Figure 6.6

In particular the equation $E = \frac{1}{2}\vec{L}(t) \cdot \vec{w}(t)$ implies that the angular velocity $\vec{w}(t)$ (having constant magnitude) makes a constant angle with the constant vector $\vec{L}$. The constant angle we denote it by θ is given by

$$\cos\theta = \frac{\vec{L} \cdot \vec{w}(t)}{\|\vec{L}\|\|\vec{w}(t)\|}.$$

Thus in the stationary frame $\mathcal{F}$ also, the angular velocity $\vec{w}(t)$ precesses about $\vec{L}$ and, thus describes a cone having the semi-angle θ; the cone having its axis along $\vec{L}$ (see Fig. 6.7(a)). Thus we see that there are two cones, one described by $\vec{w}(t)$ in the stationary frame $\mathcal{F}$ about the Z-axis ($//\vec{L}$), the other described by the same $\vec{w}(t)$ in the body frame $\hat{\mathcal{F}}$ about its $\hat{z}$-axis. Note that the two cones have $\vec{w}(t)$ as the common line along which they touch each other. Therefore, the motion of the body is such that the two cones described by $\vec{w}(t)$ roll on each other, the contact line being always along the angular vector $\vec{w}(t)$. (see Fig. 6.7(b)).

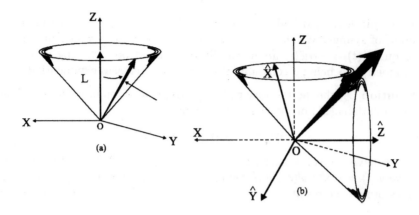

Figure 6.7

Finally, we consider a stronger case of a symmetry of a rigid body in which the body has a dynamical symmetry about a point of it in the sense of the following definition:

Definition 4 A rigid body is said to have **dynamical symmetry about a point** 0 of it if there exist a body frame $\hat{\mathcal{F}}$ with its origin at 0 with respect to which the inertia tensor has the form

$$I = \begin{bmatrix} A & 0 & 0 \\ 0 & A & 0 \\ 0 & 0 & A \end{bmatrix}$$

Thus (i) a sphere, (ii) a cube have dynamical symmetry about their geometric center (we assume here that the mass distribution in them is uniform. It can be shown that a regular tetrahedron with uniform mass distribution has dynamical symmetry about its centroid. This fact is left for the reader to prove as Exercise 17.

Referring back to *Definition 4*, using *Example 7* it can be shown that the inertia tensor has the same matrix (∗) with respect to any body frame $\hat{\mathcal{F}}$ having its origin at the point of dynamical symmetry.

Clearly the Euler's equations for such a body with dynamical symmetry about a point rotating about the same point (having dynamical symmetry) take the following form

$$A\dot{\Omega}_i(t) = N_i(t) \quad 1 \le i \le 3 \tag{23}$$

the solutions of which are

$$\Omega_i(t) = \Omega_i(o) + \frac{1}{A} \int_0^t N_i(s)ds.$$

Equivalently, $\vec{w}(t) = \vec{w}(o) + \frac{1}{A} \int_0^s \vec{N}(s)ds.$

Example 9 A rigid body with dynamical symmetry is rotating about the point of symmetry. The body is acted upon by a force which produces a torque $N(t) = -\alpha \vec{w}(t)$ about the fixed point ($\equiv$ point having dynamical symmetry), α being a constant. Prove that α decreases exponentially.

Solution Let the inertia tensor of the body be as given in ($*$) above. Then the equations (21) of rotational motion are

$$A\dot{\Omega}_i(t) = -\alpha \Omega_i(t) \quad i = 1, 2, 3.$$

which have the solutions: $\Omega_i(t) = e^{-(\frac{\alpha}{A})t}\Omega_i(o)i = 1, 2, 3$. Equivalently we have $\vec{w}(t) = e^{-(\frac{\alpha}{A})t}\vec{w}(o), \vec{w}(o)$ being the initial angular velocity of the body. This proves the result. $\square$

EXERCISES

1. The point $(a, 2a, -a), (-a, -a, a)$ and (a, a, a) of a rigid body have instantaneous velocities $(\frac{\sqrt{3}}{2}v, 0, \frac{\sqrt{3}}{2}v), (-\frac{v}{\sqrt{3}}, 0, -\frac{v}{\sqrt{3}})$ and $(0, -\frac{v}{\sqrt{3}}, \frac{v}{\sqrt{3}})$ respectively. Show that the line through the origin having direction cosines $(\frac{1}{\sqrt{3}}, \frac{-1}{\sqrt{3}}, \frac{1}{\sqrt{3}})$ is the instantaneous axis of rotation and the instantantaneous speed is $\frac{v}{2a}$.

2. A circular disc having radius a, mass M and center C is free to rotate in a vertical plane about a fixed point A on its circumference. It is given that the moment of inertia of the disc about the axis of rotation is $\frac{3}{2}Ma^2$. Let θ be the angle made by the line $A\,C$ with the vertical. Suppose the disc was released from rest when $\theta = \theta_o$. Determine the angular velocity of the disc as a function of θ.

3. Show that none of the moments of inertia can exceed the sum of the other two.

4. Prove that the sum of the diagonal terms in an inertia matrix of a right body is independent of the body frame.

5. A rigid body consists of three particles, P_1 having mass $4m$ located at the point $(a, -a, a)$, P_2 having mass $3m$ located at the point $(-a, a, a)$ and P_3 having mass $2m$ located at $(a, a, 0)$ of a body frame. Determine the inertia tensor.

6. A rigid body consists of three particles: P_1 having mass $4m$ located at $(a, -a, 0)$, P_2 having mass $3m$ located at $(-a, a, 0)$ and P_3 having mass $2m$ located at $(a, a, 0)$. Determine (i) the inertia tensor, (ii) the principal moments of inertia and (iii) the direction cosines of the principal axes.

7. Show that the angular momentum $\vec{L}(t)$ of a rigid body with respect to the center of mass is parallel to the angular velocity $\vec{w}(t)$ if and only if $\vec{w}(t)$ is parallel to a principal axis of the body.

8. A uniform cube of mass M sides $2a$ is free to rotate about its center
   0 which is held fixed. The cube is set in motion with initial velocity
   (u_1, u_2, u_3), u_i being parallel to the ith edges. Show that the kinetic
   energy of the cube is

$$\frac{M}{22}\left[5(u_1^2 + u_2^2 + u_3^2) + 6(u_1u_2 + u_2u_3 + u_3u_1)\right].$$

9. Two conical shells of semi-vertical angles β and $\beta+2\alpha$ have a common
   vertex and a common axis. Between them is pressed a solid cone
   of semi-vertical angle α. The conical shells are made to rotate in
   opposite sense with angular speed Ω about their common axis.

   (i) Determine the angular velocity of the solid cone.

   (ii) Determine its angular momentum about its vertex.

   (iii) Determine its kinetic energy.

10. A uniform disc of radius a, mass M rolls without slipping on a hor-
    izontal plane, making a constant angle β with the horizontal plane.
    Suppose the center of the disc describes a circle of radius $4a$ with
    velocity v. Determine the angular velocity and the kinetic energy of
    the disc.

11. Consider a symmetric rigid body with inertia tensor

$$I = \begin{bmatrix} A & 0 & 0 \\ 0 & A & 0 \\ 0 & 0 & B \end{bmatrix}$$

   with $B < A$. Suppose the body is free to rotate freely about the
   origin of the body frame. Describe the force free motion.

12. A thin, uniform disc of radius a, mass M and center C rolls without
    slipping on a horizontal plane, the plane of the disc making a constant
    angle with the vertical. The center moves with constant speed v in a
    horizontal circle of radius $2a$. If T denotes the kinetic energy of the
    disc, prove $\frac{3}{4} < \frac{T}{MV^2} < \frac{25}{32}$.

13. A ball of radius a is rolling without slipping on a horizontal plane,
    $\vec{w}(t)$ being its instantaneous angular velocity. Prove that the velocity
    $\vec{v}(t)$ of its center is given by $\vec{v}(t) = a\vec{w}(t) \times \vec{k}$, $\vec{k}$ being a unit vertically
    upward vector.

14. A rough and bollow sphere S of radius $3a$ is free to rotate about its
    center 0. Inside the sphere are five spheres S_0, S_1, S_2, S_3, S_4 each of
    radius a. S_0 is restricted so that its center coincides permanently with
    the center of S. The remaining spheres remain within the annular

region between S and S_0. No slipping occurs in the ensuing motion
of sphere $S_o, \cdots S_4$ under the rotational motion of S. Prove that the
spheres S_1, S_2, S_3, S_4 move in such a way that their distances remain
constant.

15. The base of a cone of height h and semi-vertical angle α rolls without
    slipping on a horizontal plane. Its vertex is fixed at a point which is
    at a height equal to the radius of the base of the cone so that the axis
    of the cone remains parallel to the horizontal plane. The axis of the
    cone makes one revolution about the fixed point in time τ. Determine
    the angular velocity of the cone.

16. Prove that the inertia tensor of a symmetric rigid body about the
    point of dynamical symmetry is independent of the orientation of the
    body frame.

17. T denotes a regular tetrahedron of uniformly distributed mass M
    and edge length a. Let $\hat{\mathcal{F}}$ be any body frame leaving its origin at the
    centroid of T, its XOY-plane parallel to a face and the z-axis passing
    through the opposite vertex. Calculate the inertia matrix of T with
    respect to $\hat{\mathcal{F}}$.

Chapter 7

Lagrangian Mechanics

Ours, according to Leibniz, is the best of all possible worlds and the laws of nature can therefore be described in terms of extremal principles.

Carl Ludwig Segal

7.1   Introduction

In this final chapter, we discuss a rudimentary account of the motion of a system of particles, the movements of the particles being subjected to a type of restrictions called **holonomic constraints**.

Let

$$\{P_1, P_2, \cdots, P_n\} \tag{1}$$

be a family of particles having masses $m_1, m_2, \cdots, m_n$ respectively. Recall, by an instantaneous *position* of (1) we mean the configuration of the particles in the physical space at an instance.

We consider the set of all the possible instantaneous positions of (1) and call it the **configuration space** of (1) and denote it by C.

Let, as usual, $\vec{r}_j(t)$ be the instantaneous position vector of the jth particle P_j. Then the instantaneous position of (1) is represented by the vector

$$\vec{r}(t) = (\vec{r_1}(t), \vec{r_2}(t), \cdots, \vec{r_n}(t))$$

in the Euclidean Space $\mathbb{R}^{3n}$. Consequently, the configuration space C is a subset of $\mathbb{R}^{3n}$. About C there are two possibilities.

Firstly, C may be an open subset of $\mathbb{R}^{3n}$, consequently if at an instant the system was at the position $(\vec{s}_1, \vec{s}_2, \cdots, \vec{s}_n)$, then each particle P_j can move a bit in *any direction* with *any speed*, from its position $\vec{s}_j$. On account of this freedom of movement, we say that the motion of (1) is *unconstrained*.

On the other hand, C may be more complicated then an open subset. In this case, there are some positions $\alpha = (\vec{s}_1, \vec{s}_2, \cdots, \vec{s}_n)$ of (1) such that at least one particle of it, say P_i, can not move in certain directions from its position $\vec{s}_i$. Thus, now there are some restrictions on the particle system in moving away from some positions like α. We express this situation by saying that the motion of the particle system is *constrained*.

Here are some illustrative examples of constrained motion

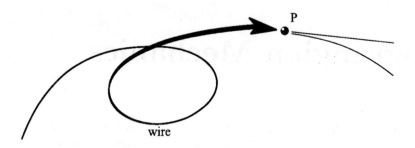

Figure 7.1

(I) A bead is sliding along a smoothly bent, thin wire. The wire represents the configurtion space of the bead (which is regarded as a particle). Note that when the bead is at a point A of the wire, it can move only in a direction tangential to the wire at the point A.(See fig 7.1)

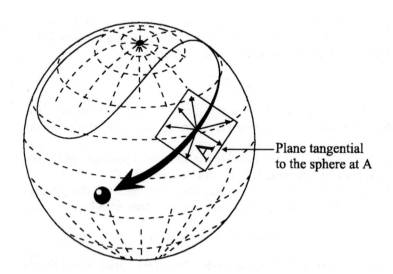

Figure 7.2

(II) When a particle is constrained to roll on a spherical surface which is its configuration space, then moving away from a point A on the sphere, it can move in a direction which is tangential to the sphere at the point A, that is the particle can have any velocity within the tangent plane to the sphere at the point A. (See fig 7.2)

(III) A more complicated example is that of a *double pendulum* shown in Fig. 7.3(a). Both the pendulae are constrained to move in a vertical plane through the point of support O. The bob B_1 describes the circle S_1. The bob B_2 describes a moving circle S_2, the bob B_1 being always at the center of S_2. Now it is difficult to imagine the configuration space of the double pendulum. But we can think of a *mathematical model* of it in the form of the Cartesian product $S_1 \times S_2$ called the *two dimensional torus* shown in Fig. 7.3(b).

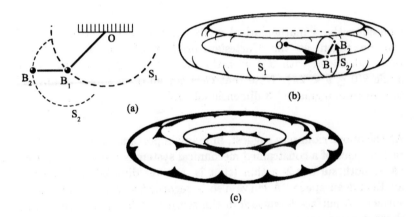

(a) (b) (c)

Figure 7.3

Note that through the motion of both the pendulae is in a plane, the mathematical model of its configuration space is a surface which is not a part of any plane. We do not mind this! What we want is that the two angles θ_1, θ_2 in (a) which specify an instantaneous position of the double pendulum determine a unique point on the torus and vice versa.

(IV) A still more complicated example is a system consisting of three particles. P_1, P_2, P_3 forming the vertices of a triangle (with fixed side lengths) and moving in a fixed plane.

Clearly, a position of the (triangular) system $\{P_1, P_2, P_3\}$ is determined completely by the following two items: (i) the position of one of the particles, say P_1, and (ii) the angle θ made by a side, say $P_1 P_2$ with a fix line (we call it L) in the plane of motion.

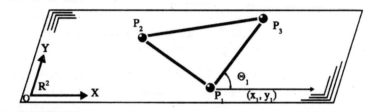

Figure 7.4

Consequently, a mathematical model for the configuration space C can be taken as the Cartesian product of the plane of motion (identified with $\mathbb{R}^2$) with the unit circle corresponding to the angle $\theta(0 \leq \theta < 2\pi)$. Note that C is not an open subset of any $\mathbb{R}^n$ but an (unimaginable) 3-dimensional surface.

As the above examples suggest, in many important cases the configuration space of a constrained mechanical system is a smooth surfaces. (A smooth surface is a thin (that is, a lower dimensional) sheet in an Euclidean space. A curve also is regarded as a one dimensional surface. A surface is *smooth* in the sense that it has no creases, no sharp corners, etc.)

Thus, a certain type of constraints on the motion of the family of particles (1) results into a geometric structure on its configuration space C, namely a smooth surface. We name this type of constraints in the following definition:

Definition 1 The constraints operating on the mechanical system (1) are said to be **holonomic** if the configuration space of its is a smooth surface.

In particular, a particle sliding along a smooth curve is said to be subjected to holonomic constraints.

Of course, smooth surfaces can be very complicated objects of geometry. Though it is a (small) digression from our subject, let us have a look at the following figures to appreciate how complicated a smooth surface in the usual 3-dimensional space can be (see Fig. 7.5) .

Now it should be clear to the reader that the more complicated a surface is, the more complicated must be the nature of the motion on it. We will therefore consider a type of surfaces which are simple enough. Before explaining the sense in which they are simple enough, let us examine the role of Cartesian coordinates which we use to parametrise the positions of a mechanical system.

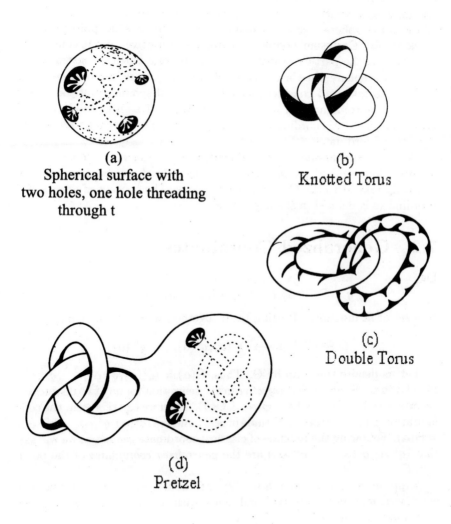

(a)
Spherical surface with
two holes, one hole threading
through t

(b)
Knotted Torus

(c)
Double Torus

(d)
Pretzel

Figure 7.5   Surfaces

When C is an open subset of an Euclidean space $\mathbb{R}^{3n}$, we need all the $3n$ coordinates $(x^1, y^1, z^1, x^2, y^2, z^2, \cdots, x^n, y^n, z^n)$. But when the motion is constrained, there are many cases in which the coordinates are not independent of each other. Interdependence between the coordinates gives rise to inter-dependence between the equations of motion. We may try to drop the dependent coordinates, but it turns out that it is possible to do so only on small patches and *not* on the whole of the configuration space. For example, on the sphere $x^2 + y^2 + z^2 = 1$, we may drop the z coordinate which depends on the coordinates x and $y : z = \pm\sqrt{(1 - x^2 - y^2)}$, only on the upper semi sphere or on the lower semi-sphere, (but not on both that is on the whole sphere, because, the ambiguity in the $\pm$ sign spoils the act.)

Secondly, Cartesian coordinates are not sensitive to the constraints operating on the particle system. Thus in case of a particle constrained to move on the sphere, the Cartesian coordinates x, y, z are (not only excessive but) fail to take into consideration, the spherical symmetry of the sphere (which is the configuration space of the particle.) This is precisely why we prefer the spherical coordinates (that is, the lattitude and longitude) to the Cartesian coordinates.

We therefore assume that the configuration space is simple enough to admit independent parameters called *generalized coordinates* which are smoothly related to the Cartesian coordinates. The concept of generalized coordinates is defined in the next section.

7.2   Generalized Coordinates

Let

$$q^1 : C \longrightarrow \mathbb{R}; q^2 : C \longrightarrow \mathbb{R}; \cdots; q^N : C \longrightarrow \mathbb{R} \qquad (2)$$

be a set of m functions. Putting them together, we get a single map

$$C \longrightarrow \mathbb{R}^N; \alpha \longmapsto (q^1(\alpha), q^2(\alpha), \cdots, q^m(\alpha)) \qquad (3)$$

Let us denote this map by Θ. Thus, $\Theta(\alpha) = (q^1(\alpha), q^2(\alpha), \cdots, q^m(\alpha))$ for all α in C. Suppose Θ is *injective* in the sense that for any two distinct elements α and β of C, we have $\Theta(\alpha) \neq \Theta(\beta)$. Then we can represent the configuration α by the ordered N-tuple of real numbers $(q^1(\alpha), q^2(\alpha), \cdots q^N(\alpha))$ $= \Theta(\alpha)$. Following the practice of classical coordinate geometry, we will say that $(q^1(\alpha), q^2(\alpha), \cdots, q^N(\alpha))$ are the *generalized coordinates* of the point α in C.

Suppose, we have chosen a set (2) of functions in such a way that the associated map Θ is injective (and hence each $\alpha \in C$ has the generalized coordinates $(q^1(\alpha), q^2(\alpha), \cdots q^N(\alpha))$.

Now there are tow sets of coordinates for each $\alpha \in C$: (1) The usual Cartesian coordinates:

$$(x^1(\alpha), y^1(\alpha), z^1(\alpha), \cdots, x^n(\alpha), y^n(\alpha), z^n(\alpha)).$$

and (2) the generalized coordinates:

$$(q^1(\alpha)q^2(\alpha)\cdots q^N(\alpha)).$$

Now, because Θ is assumed to be injective, the Cartesian coordinates become functions of the generalized coordinates:

$$\left.\begin{array}{rcl} x^i(\alpha) & = & x^i(q^1(\alpha),\quad q^2(\alpha),\quad \cdots \quad q^N(\alpha)) \\ y^i(\alpha) & = & y^i(q^1(\alpha),\quad q^2(\alpha),\quad \cdots \quad q^N(\alpha)) \\ z^i(\alpha) & = & z^i(q^1(\alpha),\quad q^2(\alpha),\quad \cdots \quad q^N(\alpha)) \end{array}\right\} \quad 1 \leq i \leq n.$$

for all $\alpha \in C$. In other words we can use Θ^{-1} and express the Cartesian coordinates as the functions of the generalized coordinates:

$$\left.\begin{array}{rcl} x^i & = & x^i(q^1, q^2, \cdots, q^N) \\ y^i & = & y^i(q^1, q^2, \cdots, q^N) \\ z^i & = & z^i(q^1, q^2, \cdots, q^N) \end{array}\right\} \tag{4}$$

We assume that the functions (2) satisfy the following two additional properties:

(A) The functions (4) (i.e.,the Cartesian coordinates as functions of the generalized coordinates $q^1, q^2, \cdots, q^N$) are smooth functions.

(B) The functions (2) are *independent* in the sense that there is no smooth function $F : \mathbb{R}^N \longrightarrow R$ satisfying

$$F(q^1, q^2, \cdots, q^N) \equiv 0.$$

This discussion leads to the following definition:

Definition 2 A set of **generalized coordinates** (**generalized coordinate functions,** to be more accurate) is a set of functions (2) satisfying the following conditions:

(I) The map; $C \longrightarrow \mathbb{R}^N; \alpha \longmapsto (q^1(\alpha), q^2(\alpha) \cdots, q^N(\alpha))$ is injective.

(II) The functions described in (4) are smooth.

(III) The functions (2) are independent.

Of course, when the configuration space admits one set $(q^1, q^2, \cdots, q^N)$ of generalized coordinates, it will admit many more. But it is an elementary result (which we do not prove) that all such coordinate sets contain the same number of coordinate functions.

This common integer, namely N here - is an important dynamical attribute of the constraints operating on the particle system. It is called the **number of degrees of freedom** of the particle system.

Thus, in the illustrative example (I) in Chapter 1, let $\ell(A)$ be the length of a point A on the wire from one end of it. Then $\ell : C \longrightarrow [0, L]$ is a (single) generalized coordinate of the bead and hence the bead has one degree of freedom.

In example (II), the angles describing lattitude and longitude represent a set of generalized coordinates. Therefore, the particle rolling on the spherical surface has two degrees of freedom.

For the double pendulum of example (III) the two angles (θ_1, θ_2) serve as a set of generalized coordinates and so, the double pendulum is a constrained mechanical system having two degrees of freedom.

The system described in example (IV) has three degrees of freedom, because the coordinates (x, y, θ) described there are a set of generalized coordinates.

7.3   Motion

We continue with our discussion of the motion of the system (1) which is subjected to holonomic constraints having N degrees of freedom.

We choose a set $(q^1, q^2, \cdots, q^N)$ of generalized coordinates on its configuration space C. From now-onwards, in order to shorten our notation, we write simply q for the N-tuple $(q^1, q^2, \cdots, q^N)$ which are had denoted by Θ earlier. Thus, q stands for the map
$$C \longrightarrow \mathbb{R}^N; \alpha \longmapsto \big(q^1(\alpha), q^2(\alpha), \cdots, q^N(\alpha)\big).$$

As usual, the motion of the particle system is described by means of a smooth curve
$$c : I \longrightarrow C \tag{5}$$

which we have been calling the trajectory of (1), I being a time interval and for each $t \in I, c(t)$ being the position of (1) at the instant t.

We now use the chosen generalized coordinates to express the instantaneous position $c(t)$ as an ordered N-tuple of real numbers, the instantaneous generalized coordinates of $c(t)$

$$\big(q^1(c(t)), q^2(c(t)), \cdots, q^N(c(t))\big)$$

which we will write more simply as $\big(q^1(t), q^2(t), \cdots, q^N(t)\big)$. Thus the trajectory (5) is expressible in terms of N real functions of the time variable t

$$t \longmapsto q^1(t); t \longmapsto q^2(t), \cdots, t \longmapsto q^N(t).$$

Now our standing assumption that the motion is smooth clearly implies that these functions are smooth functions of time. We differentiate them to get the components (with respect to the chosen generalized coordinates) of the instantaneous velocity of (1).

Definition 2 The derivatives $\big(\dot{q}^1(t), \dot{q}^2(t), \cdots, \dot{q}^N(t)\big)$ are the **generalized velocities of** (1) at the instant t.

$\dot{q}^i(t)$ is said to be the generalized velocity component *conjugate to* the coordinate $q^i(t)$.

Just as in case of coordinates, we abbreviate the notation by writing $\dot{q}(t)$ for $\left(\dot{q}^1(t), \dot{q}^2(t), \cdots, \dot{q}^N(t)\right)$.

7.4   The Lagrangian Function

Recall, the kinetic energy of (1) is given by

$$T(t) = \frac{1}{2} \sum_{\ell=1}^{n} m_\ell \|\dot{\vec{r}}_\ell(t)\|^2.$$

Now $\vec{r}_\ell(t) = \vec{r}_\ell(q^1(t), q^2(t), \cdots, q^N(t))$ gives (on differentiation) $\dot{\vec{r}}_\ell(t) = \sum_{i=1}^{N} \frac{\partial \vec{r}_\ell}{\partial q^i}(q(t))\dot{q}^i(t)$ and consequently,

$$
\begin{aligned}
T(t) &= \frac{1}{2} \sum_{\ell=1}^{n} m_\ell \sum_{i,j=1}^{N} \frac{\partial \vec{r}_\ell}{\partial q^i}(q(t)) \cdot \frac{\partial \vec{r}_\ell}{\partial q^j}(q(t))\dot{q}^i(t)\dot{q}^j(t). \\
&= \frac{1}{2} \sum_{i,j=1}^{N} \left\{ \sum_{\ell=1}^{n} m_\ell \frac{\partial \vec{r}_\ell}{\partial q^i}(q(t)) \cdot \frac{\partial \vec{r}_\ell}{\partial q^j}(q(t)) \right\} \dot{q}^i(t)\dot{q}^j(t). \\
&= \frac{1}{2} \sum_{i,j=1}^{N} a_{ij}(q(t))\dot{q}^i(t)\dot{q}^j(t).
\end{aligned}
$$

where we have written $a_{ij}(q(t))$ for the sum $\sum_{\ell=1}^{n} m_\ell \frac{\partial \vec{r}_\ell}{\partial q^i}(q(t)) \cdot \frac{\partial \vec{r}_\ell}{\partial q^j}(q(t))$. Thus, we have the smooth functions $a_{ij} : C \longrightarrow \mathbb{R}, 1 \leq i, j \leq N$ with $a_{ij} = a_{ji}$ such that

$$T = \frac{1}{2} \sum_{i,j=1}^{N} a_{ij}(q(t))\dot{q}^i(t)\dot{q}^j(t) = T(q(t), \dot{q}(t)).$$

Also, note that $T(q(t), \dot{q}(t)) \geq 0$ and for any $q(= q(t)) \in C, T(q(t), \dot{q}(t)) = 0$ for all $\dot{q}(t)$ implies $a_{ij}(q) = 0$, we get that $[a_{ij}(q)]$ is a *positive definite matrix* for each $q \in C$.

Next, we consider the potential energy function $U : C \longrightarrow \mathbb{R}$ and also, the difference $T - U$ where we are treating it as a function of both q and $\dot{q}$. We denote it by L. Thus,

$$L = T - U = \frac{1}{2} \sum a_{ij}(q)\dot{q}^i\dot{q}^j - U(q) = L(q, \dot{q}).$$

Definition 3 $L(q, \dot{q})$ is called the Lagrangian function of the (conservative) mechanical system (1).

Since we prefer the generalized coordinates to the Cartesian coordinates, we need obtain equations of motion for the trajectory (5) in terms of the generalized coordinates $q = (q^1, q^2, \cdots, q^N)$ and their time derivatives. We will differentiate the Lagrangian function partially with respect to the q^i and the $\dot{q}^j$ and obtain out of these derivatives a certain type of second order differential equations (as many as there are the degrees of freedom). These equations, called Lagrange's equations will be the desired equations of motion and will be dealt in the next section.

In the rest of this section, we consider some (constrained) mechanical systems and obtain their Lagrange functions. Of course, the Lagrangian functions will depend on the choice of generalized coordinates.

Example 1 A mechanical system called *double pendulum* was introduced in the **Introduction** section of this chapter (see Fig. 7.1(a)). A double pendulum consists of two pendulae P_1 and P_2 joined together by hanging second pendulum from the bob of the first pendulum as shown in the figure.

The bobs B_1 and B_2 have masses m_1 and m_2 respectively and the rods, which have negligible masses have lengths ℓ_1 and ℓ_2 respectively. Taking the angles θ_1 and θ_2 (as shown in the figure) as the generalized coordinates obtain the Lagrangian function of the double pendulum.

Solution Let $(x_1(t), y_1(t))$ be the Cartesian coordinates of the bob B_1 and let $x_2(t), y_2(t))$ be those of the bob B_2. We use the angles $\{\theta_1, \theta_2\}$ as the generalized coordinates. Then the Cartesian coordinates of the bobs are related to the generalized coordinates by the equations

$$\left. \begin{array}{llll} x_1(t) & = & \ell_1 \cos\theta_1(t), & x_2(t) & = & \ell_1 \cos\theta_1(t) & + & \ell_2 \cos\theta_2(t). \\ y_1(t) & = & \ell_1 \sin\theta_1(t), & y_2(t) & = & \ell_1 \sin\theta_1(t) & + & \ell_2 \cos\theta_2(t). \end{array} \right\} \quad (*)$$

Differentiating equations $(*)$ with respect to time, we get

$$\dot{x}_1 = -\ell_1 \dot{\theta}_1 \sin\theta_1 ; \dot{y}_1 = \ell_1 \dot{\theta}_1 \cos\theta_1 \qquad \text{and}$$

$$\dot{x}_2 = -\ell_1 \dot{\theta}_1 \sin\theta_1 - \ell_2 \dot{\theta}_2 \cos\theta_2 ; \dot{y}_2 = \ell_1 \dot{\theta} \cos\theta_1 + \ell_2 \dot{\theta}_2 \cos\theta_2.$$

Hence the kinetic energy T of the double pendulum is:

$$
\begin{aligned}
T & = \frac{m_1}{2} \left[\dot{x}_1^2 + \dot{y}_1^2 \right] + \frac{m_2}{2} \left[\dot{x}_2^2 + \dot{y}_2^2 \right] \\
& = \frac{m_1}{2} \left[\ell_1^2 \dot{\theta}_1^2 \sin^2\theta_1 + \ell_1^2 \dot{\theta}_1^2 \cos^2\theta_1 \right] \\
& \quad + \frac{m_2}{2} \left[(-\ell_1 \dot{\theta}_1 \sin\theta_1 - \ell_2 \dot{\theta}_2 \cos\theta_2)^2 + (\ell_1 \dot{\theta}_1 \cos\theta_1 + \ell_2 \dot{\theta}_2 \cos\theta_2^2) \right] \\
& = \frac{m_1}{2} \ell_1^2 \dot{\theta}_1^2 + \frac{m_2}{2} \left[\ell_1^2 \dot{\theta}_1^2 + \ell_2^2 \dot{\theta}_2^2 + 2\ell_1 \ell_2 \dot{\theta}_1 \dot{\theta}_2 (\sin\theta_1 \sin\theta_2 + \cos\theta_1 \cos\theta_2) \right] \\
& = \frac{m_1}{2} \ell_1^2 \dot{\theta}_1^2 + \frac{m_2}{2} \left[\ell_1^2 \dot{\theta}_1^2 + \ell_2^2 \dot{\theta}_2^2 + 2\ell_1 \ell_2 \dot{\theta}_1 \dot{\theta}_2 \cos(\theta_1 - \theta_2) \right] \\
& = \frac{m_1 + m_2}{2} \ell_1^2 \dot{\theta}_1^2 + \frac{m_2}{2} \ell_2^2 \dot{\theta}_2^2 + \ell_1 \ell_2 m_2 \dot{\theta}_1 \dot{\theta}_2 \cos(\theta_1 - \theta_2).
\end{aligned}
$$

Clearly, potential energy of the double pendulum is the sum of potential energies of the two bobs:

$$
\begin{aligned}
U &= m_1 g y_1(t) + m_2 g y_2(t) = m_1 g(\ell_1 \sin\theta_1) + m_2 g(\ell_1 \sin\theta_1 + \ell_2 \sin\theta_2) \\
&= U(\theta_1, \theta_2).
\end{aligned}
$$

Therefore, the Lagrangian function of the double pendulum is

$$
\begin{aligned}
L &= T - U \\
&= \frac{(m_1 + m_2)}{2} \ell_1^2 \dot{\theta}_1^2 + \frac{m_2}{2} \ell_2^2 \dot{\theta}_2^2 + 2\ell_1 \ell_2 m_2 \dot{\theta}_1 \dot{\theta}_2 \cos(\theta_1 - \theta_2) \\
&\quad - m_1 g \ell_1 \sin\theta_1 - m_2 g(\ell_1 \sin\theta_1 + \ell_2 \sin\theta_2).
\end{aligned}
$$

$\square$

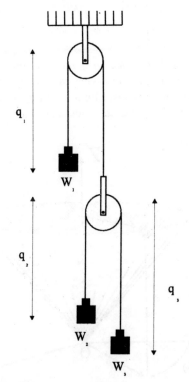

Double Atwood machine

Figure 7.6

Example 2 (The Double Atwood Machine) It consists of three masses m_1, m_2 and m_3 considered as particles hanging from weight less pullies P_1, P_2 by means of light ropes R_1, R_2 having inextensible lengths ℓ_1 and ℓ_2. The whole system remains in a vertical plane and the masses are constrained to move only upwards or downwards. The arrangement is shown in Fig. 7.6

Solution We take the generalized coordinates q_1 and q_2 as shown in the figure. Then we have

$$
\begin{aligned}
T &= \frac{W_2}{2}\dot{q}_1^2 + \frac{W_2}{2}(\ell_1 - q_1 + q_2)^{\dot{2}} + \frac{W_3}{2}(\ell_1 - q_1 + \ell_2 - q_2)^{\dot{2}} \\
&= \frac{W_1}{2}\dot{q}_1^2 + \frac{W_2}{2}(\dot{q}_1 - \dot{q}_2)^2 + \frac{W_3}{2}(\dot{q}_1 + \dot{q}_2)^2 \quad \text{and} \\
U &= -W_1 g q_1 - W_2(\ell_1 - q_1 + q_2)g - W_3(\ell_1 - q_1 + \ell_2 - q_2)g.
\end{aligned}
$$

Hence, the Lagrangian function of the Atwood's machine is given by:

$$
\begin{aligned}
L &= L(q_1, q_2, \dot{q}_1, \dot{q}_2) \\
&= \frac{W_1}{2}\dot{q}_1^2 + \frac{W_2}{2}(\dot{q}_1 - \dot{q}_2)^2 + \frac{W_3}{2}(\dot{q}_1 + \dot{q}_2)^2 \\
&\quad + [W_1 q_1 + W_2(\ell_1 - q_1 + q_2) + W_3(\ell_1 - q_1 + \ell_2 - q_2)]\, g \qquad \square
\end{aligned}
$$

Example 3 A particle P of mass m is moving under gravity on a smooth conical surface. The cone has semi-vertical angle α and it is held with its axis vertical and apex down. Describe a set of generalized coordinates and using them, obtain the Lagrangian function of the particle.

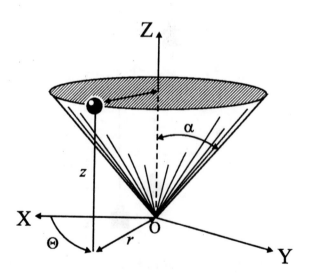

Figure 7.7

Solution We take the generalized coordinates (r, θ) as shown in the Fig.7.7. Let (x, y, z) be the Cartesian coordinates of the particle. Then we have

$$
x = r\cos\theta, \qquad y = r\sin\theta, \qquad z = r\cot\alpha.
$$

Therefore, their time derivatives are given by

$$\dot{x} = \dot{r}\cos\theta - r\dot{\theta}\sin\theta, \dot{y} = \dot{r}\sin\theta + r\dot{\theta}\cos\theta, \text{ and } \dot{z} = \dot{r}\cot\alpha.$$

The kinetic energy of the particle is now given by

$$T = \frac{m}{2}\left[\dot{x}^2 + \dot{y}^2 + \dot{z}^2\right] = \frac{m}{2}\left[\dot{r}^2\csc^2\alpha + r^2\dot{\theta}^2\right].$$

The potential energy of the particle is given by $U = mgz = mgr\cot\alpha$. Therefore, the Lagrangian of the particle is

$$L = T - U = \frac{m}{2}\left[\dot{r}^2\csc^2\alpha + r^2\dot{\theta}^2\right] - mgr\cot\alpha.$$

□

7.5 Lagrange's Equations

Recall, the equations $m_\ell\ddot{\vec{r_\ell}} = F_\ell; 1 \le \ell \le n$ govern the motion of the particles in the system (1). We considered these equations in **Chapter 5** where the particles were moving in their domains of motion, the particle P_ℓ being acted upon by an *applied force* field $F_\ell = F_\ell^{(e)} + \sum\limits_{j(j\neq\ell)} F_{\ell j}$ but there were no constraints on the motion.

When the system (1) is subjected to holonomic constraints, there are additional forces, namely the **forces of constraints**. These are the forces which prevent the particles from comming out of the configuration space even when the external and the internal forces may pull the particles out. Consequently, the equation of motion now takes the form

$$m_\ell\ddot{r_\ell} = F_\ell^{(a)} + F_\ell^{(c)}$$

where we have written $F_\ell^{(a)}$ for $F_\ell + \sum\limits_{j(j\neq\ell)} F_{\ell j}$ and $F_\ell^{(c)}$ is the field of forces of constraints operating on P_ℓ.

We consider an arbitrary configuration $\alpha = (\vec{s_1}, \vec{s_2}, \cdots, \vec{s_n})$ of the system (1). The forces of constraint operating on (1) at α can be represented by means of a single vector in $\mathbb{R}^{3n}$:

$$\left(\vec{F}_1^{(c)}(\alpha), \vec{F}_2^{(c)}(\alpha), \cdots, \vec{F}_n^{(c)}(\alpha)\right) \tag{6}.$$

In general, the vector (6) depends on the following two aspects:

(i) The geometry of the configuration space.

(ii) The state of motion of the system at the configuration α.

This dependence of the forces of constraint on (i) and (ii) can be very complicated but a simple geometric principle (stated below) enables us to eliminate the forces of constraint altogether from the equations of motion.

D'Alembert's Principle The forces of constraint, that is, the vector (6) is perpendicular to the configuration C at α.

The logic behind D'Alembert's principle is simple enough. Indeed, the forces of constraint are supposed to prevent the system (1) from comming *out of* the configuration space. On the other hand the only physical agency trying to pull the system (1) out of C (from the place α) is the normal component of the (consolidated) force $(\vec{F}_1^{(a)}(s_1), \vec{F}_2^{(a)}(s_2), \cdots, \vec{F}_n^{(a)}(s_n))$. Consequently, the force (6) must be (equal in magnitude and) opposite to this normal component. In particular (6) must be normal to C at α and hence the D'Alembert's principle.

Now, we consider an infinitesimal displacement $(d\vec{r}_1, d\vec{r}_2, \cdots, d\vec{r}_n)$ of the system (1) from a configuration α. Thus in the infinitesimal displacement of the system, the particles P_j undergoes a small displacement $d\vec{r}_j$ from the position $\vec{s}_j$, this displacement being in accordance with the constraints on the system. Consequently, at the infinitesimal level, the displacement is tangential to configuration space C at α. Now, the perpendicularity relation stipulated in D'Alembert's principle implies

$$\left(\vec{F}_1^{(c)}(s_1), \vec{F}_2^{(c)}(s_2), \cdots, \vec{F}_n^{(c)}(s_n) \right) \cdot (d\vec{r}_1, d\vec{r}_2, \cdots, d\vec{r}_n) = 0. \qquad (7)$$

Next, we consider the equations of motion

$$m_\ell \ddot{\vec{r}}_\ell = F_\ell^{(a)} + F_\ell^{(c)}.$$

Multiplying these equations by the infinitesimal displacements $d\vec{r}_\ell$ and summing them all; we get

$$\begin{aligned} \sum_\ell m_\ell \ddot{\vec{r}}_\ell \cdot d\vec{r}_\ell &= \sum_\ell F_\ell^{(a)} \cdot d\vec{r}_\ell + \sum_\ell F_\ell^{(c)} \cdot d\vec{r}_\ell \\ &= \sum_\ell F_\ell^{(a)} \cdot d\vec{r}_\ell + 0 \quad \text{by (7)} \end{aligned}$$

Thus, we have the (infinitesimal) equation

$$\sum_\ell m_\ell \ddot{\vec{r}}_\ell \cdot d\vec{r}_\ell = \sum F_\ell^{(a)} \cdot d\vec{r}_\ell \qquad (8)$$

Now $\vec{r}_\ell = \vec{r}_\ell(q^1, q^2, \cdots, q^N)$ and therefore, we get

$$d\vec{r}_\ell = \sum_{i=1}^N \frac{\partial \vec{r}_\ell}{\partial q^i} dq^i.$$

Substituting these expressions in (8), we get

$$\sum_{i=1}^{N} \left\{ \sum_{\ell=1}^{N} m_\ell \ddot{\vec{r}}_\ell \cdot \frac{\partial \vec{r}_\ell}{\partial q^i} \right\} dq^i = \sum_{i=1}^{N} \left\{ \sum_{\ell} F_\ell^{(a)} \cdot \frac{\partial \vec{r}_\ell}{\partial q^i} \right\} dq^i.$$

Since the generalized coordinates are independent, so are their infinitesimals $\{dq^1, dq^2, \cdots, dq^N\}$ and so, equating the coefficients of dq^i on both sides of the last infinite
simal equation, we get

$$\sum_\ell m_\ell \ddot{\vec{r}}_\ell \frac{\partial \vec{r}_\ell}{\partial q^i} = \sum_\ell F_\ell^{(a)} \cdot \frac{\partial \vec{r}_\ell}{\partial q^i}; 1 \le i \le N \tag{9}$$

In this chapter, we are considering the case in which the applied forces are conservative. Let $U : C \longrightarrow \mathbb{R}$ be a potential energy function for the forces. Then $F_\ell^{(a)} = - \text{grad}_{\vec{r}_\ell} U$ and consequently

$$\sum_\ell F_\ell \cdot \frac{d\vec{r}_\ell}{\partial q^i} = - \sum_\ell \text{grad}_{\vec{r}_\ell} U \cdot \frac{\partial \vec{r}_\ell}{\partial q^i} = - \frac{\partial U}{\partial q^i}.$$

Differentiating the function $\vec{r}_\ell(q) = \vec{r}_\ell(q^1, q^2, \cdots q^N)$ with respect to time, we get

$$\dot{\vec{r}}_\ell = \sum_{j=1}^{N} \frac{\partial \vec{r}_\ell}{\partial q^j} \dot{q}^i \tag{10}$$

and now, differentiating (10) partially with respect to $\dot{q}^i$, we get

$$\frac{\partial \vec{r}_\ell}{\partial q^i} = \frac{\partial \dot{\vec{r}}_\ell}{\partial \dot{q}^i} \tag{11}$$

Now,

$$\begin{aligned}
\sum_\ell m_\ell \ddot{\vec{r}}_\ell \frac{\partial \vec{r}_\ell}{\partial q^i} &= \sum_\ell m_\ell \ddot{\vec{r}}_\ell \cdot \frac{\partial \dot{\vec{r}}_\ell}{\partial \dot{q}^i}, \text{ using (11)} \\
&= \sum_\ell m_\ell \frac{d}{dt}(\dot{\vec{r}}_\ell) \cdot \frac{\partial \dot{\vec{r}}_\ell}{\partial \dot{q}^i} \\
&= \sum_\ell m_\ell \frac{d}{dt} \left(\dot{\vec{r}}_\ell \cdot \frac{\partial \dot{\vec{r}}_\ell}{\partial \dot{q}^i} \right) - \sum_\ell m_\ell \dot{\vec{r}}_\ell \frac{d}{dt} \left(\frac{\partial \dot{\vec{r}}_\ell}{\partial \dot{q}^i} \right) \\
&= \sum_\ell m_\ell \frac{d}{dt} \frac{\partial}{\partial \dot{q}^i} \left(\frac{\dot{\vec{r}}_\ell \cdot \dot{\vec{r}}_\ell}{2} \right) - \sum_\ell m_\ell \dot{\vec{r}}_\ell \frac{d}{dt} \left(\frac{\partial \dot{\vec{r}}_\ell}{\partial q^i} \right)
\end{aligned}$$

where we have used (11) again to reinstall $\frac{\partial \vec{r}_\ell}{\partial \dot{q}^i}$ in the last factor of the second summation above.

$$
= \frac{d}{dt} \frac{\partial}{\partial \dot{q}^i} \left(\sum_\ell m_\ell \left(\frac{\dot{\vec{r}}_\ell \cdot \dot{\vec{r}}_\ell}{2} \right) \right) - \sum_\ell m_\ell \dot{\vec{r}}_\ell \frac{\partial}{\partial q^i} \frac{d}{dt} \vec{r}_\ell
$$

$$
= \frac{d}{dt} \frac{\partial}{\partial \dot{q}^i}(T) - \sum_\ell m_\ell \frac{\partial}{\partial q^i} \left(\frac{\dot{\vec{r}}_\ell \cdot \dot{\vec{r}}_\ell}{2} \right)
$$

$$
= \frac{d}{dt} \frac{\partial}{\partial \dot{q}^i}(T) - \frac{\partial}{\partial q^i} \left(\sum_\ell m_\ell \frac{\dot{\vec{r}}_\ell \cdot \dot{\vec{r}}_\ell}{2} \right)
$$

$$
= \frac{d}{dt} \frac{\partial}{\partial \dot{q}^i}(T) - \frac{\partial T}{\partial q^i}.
$$

Thus, equation $\sum_\ell F_\ell \cdot \frac{\partial \vec{r}_\ell}{\partial q^i} = -\frac{\partial U}{\partial q^i}$ leads to the equation

$$
\frac{d}{dt} \frac{\partial T}{\partial \dot{q}^i} - \frac{\partial T}{\partial q^i} = -\frac{\partial U}{\partial q^i} \tag{12}
$$

Note that the potential energy function U is a function of $q = (q^1, q^2, \cdots, q^N)$ and is independent of $\dot{q} = (\dot{q}^1, \dot{q}^2, \cdots, \dot{q}^N)$ and therefore, we have $\frac{\partial U}{\partial \dot{q}^i} = 0, 1 \le i \le N$. Consequently, equation (12) can be rewritten in the form

$$
\frac{d}{dt} \left(\frac{\partial T}{\partial \dot{q}^i} - \frac{\partial T}{\partial q^i} \right) = -\frac{\partial U}{\partial q^i} + 0 = -\frac{\partial U}{\partial q^i} + \frac{d}{dt} \frac{\partial U}{\partial \dot{q}^i} \tag{13}
$$

or equivalently, $\frac{d}{dt} \frac{\partial T}{\partial \dot{q}^i} - \frac{d}{dt} \frac{\partial U}{\partial \dot{q}^i} = \frac{\partial T}{\partial q^i} - \frac{\partial U}{\partial q^i}$ and so, in terms of the Lagrangian function of the system (1) introduced in the preceding section. Thus, we have

$$
\begin{aligned}
\frac{d}{dt} \frac{\partial L}{\partial \dot{q}^i} &= \frac{d}{dt} \frac{\partial T}{\partial \dot{q}^i} - \frac{d}{dt} \frac{\partial U}{\partial \dot{q}^i}, \quad L = T - U \\
&= \frac{d}{dt} \frac{\partial T}{\partial \dot{q}^i} - 0, \text{ since } \frac{\partial U}{\partial \dot{q}^i} = 0 \\
&= \frac{\partial T}{\partial q^i} - \frac{\partial U}{\partial q^i} \text{ by equation (13)} \\
&= \frac{\partial L}{\partial q^i}.
\end{aligned}
$$

We have now proved that the motion of the particle system (1) is governed by the second order differential equations

$$
\frac{d}{dt} \frac{\partial L}{\partial \dot{q}^i} - \frac{\partial L}{\partial q^i} = 0, 1 \le i \le N \tag{14}
$$

These differential equations are called **Lagrange's equations of motion** or sometimes, **Euler-Lagrange equations of motion**.

Before proceeding further, let us consider once again the particle system (1) with no constraints acting on it. Then we can consider the Cartesian coordinates $(x^1 y^1, z^1, x^2, y^2, z^2, \cdots x^n, y^n, z^n)$ as a particular choice of the generalized coordinates $(q^1, q^2, \cdots, q^N)$. Now, the kinetic energy of the system has the form

$$T = \frac{1}{2} \sum_{\ell} m_\ell \left[\dot{x}_\ell^2 + \dot{y}_\ell^2 + \dot{z}_\ell^2 \right]$$

while the potential energy is a smooth function $U(x^1, y^1, z^1, x^2, y^2, z^2, \cdots x^n, y^n, z^n)$ of the Cartesian coordinates. Now, we have

$$
\begin{aligned}
L &= L(x^1, y^1, z^1, \cdots x^n, y^n, z^n, \dot{x}^1, \dot{y}^1, \dot{z}^1, \cdots \dot{x}^n, \dot{y}^n, \dot{z}^n) \\
&= \frac{1}{2} \sum_\ell m_\ell \left[(\dot{x}^\ell)^2 + (\dot{y}^\ell)^2 + (\dot{z}^\ell)^2 \right] - U(x^1, y^1, z^1, \cdots x^n, y^n, z^n)
\end{aligned}
$$

Now the Lagrangian equation in the coordinate are $\frac{d}{dt} \frac{\partial L}{\partial \dot{x}^k} - \frac{\partial L}{\partial x^k} = 0$.

But $\frac{d}{dt} \frac{\partial L}{\partial \dot{x}^k} = \frac{d}{dt}(m_k \dot{x}_k) = m_k \ddot{x}^k$ and $\frac{\partial L}{\partial x^k} = -\frac{\partial U}{\partial x^k}$. Consequently, the above equation becomes $m_k \ddot{x}_k + \frac{\partial U}{\partial x^k} = 0$. or equivalently

$$m_k \ddot{x}_k = -\frac{\partial U}{\partial x^k} \tag{15}$$

Similarly, the Lagrange's equations in other coordinates are

$$m_k \ddot{y}_k = -\frac{\partial U}{\partial y^k} \tag{16}$$

$$m_k \ddot{z}_k = -\frac{\partial U}{\partial z^k} \tag{17}$$

Combining equations (15), (16) and (17) in a single vector equation, we get the usual Newtons equations

$$m_\ell \ddot{\vec{r}}_\ell = - \operatorname{grad}_{\vec{r}_\ell} U.$$

It is thus proved that Lagrange's equations generalise the familiar Newtons equations of motion. When there were no constraints, we could make use of the classical notion of *position vector* of a particle and formulate the equation of motion in terms of the time derivatives of the position vectors of the particles and the known applied force fields operating in the region. But when there are constraints, they give rise to additional forces which depend on the motion itself. Besides, we need make a more careful choice of parameters, that is, the *generalized coordinates* to specify the instantaneous positions of the mechanical system. These generalized coordinates are often non-linear in nature (e.g. the angular coordinates). Once a choice of generalized coordinates is made, Newton's equations expressed in terms of the generalized coordinates become the Lagrange's equations of motion.

7.6 Hamilton's Principle of Least Action

In the last section we used D'alembert's principle which pertained to the forces of constraints and obtained Lagrange's equations of motion. In this section, we introduce another principle-Hamilton's principle of least action and obtain the same equations from it. This new principle has immensely wide applicability throughout Natural Sciences. It is considered so much important that in Mathematics proper, it has given rise to a separate topic called **Calculus of Variations**. Though we have already obtained the equations of motion,(namely Lagrange's equations) and could go ahead with our analysis of motion, we have preferred to look at the motion from yet another point of view (which motivates Hamilton's principle) and arrive at the same equations once again.

First we describe the situation

Let A and B be two positions of the mechanical system, that is, two points of its configuration space C. Suppose, the mechanical system is to start off from A at time $t = a$ and is required to reach the place B at time $t = b$. The question that naturally arises is: Among all the possible paths in C joining A to B, along which path the mechanical system will travel in the stipulated time interval $[a, b]$?

To answer this question, we need consider the set of all smooth curves

$$\alpha : [a, b] \longrightarrow C$$

with $\alpha(a) = A$ and $\alpha(b) = B$. Let us denote this set by $C([ab])$. Thus, $C([ab])$ is the set of all trajectories available to the mechanical system enabling it to start moving from A at time $t = a$ and reach the destination B at time $t = b$.

With any such $\alpha \in C([a, b])$, we associate the real number $A(\alpha)$ defined below

$$A(\alpha) = \int_a^b L(\alpha(t), \dot{\alpha}(t)) dt.$$

Note that if $q = (q^1, q^2, \cdots, q^N)$ is a set of generalized coordinates on C, then along the curve α, we have $\alpha(t) = (q^1(t), q^2(t), \cdots, q^N(t))$ and $\dot{\alpha}(t) = (\dot{q}^1(t), \dot{q}^2(t), \cdots, \dot{q}^N(t))$. Consequently, $A(\alpha)$ is given by

$$A(\alpha) = \int_a^b L\left(q^1(t), q^2(t), \cdots, q^N(t), \dot{q}^1(t), \dot{q}^2(t), \cdots, \dot{q}^N(t)\right) dt.$$

Definition3 $A(\alpha)$ is called the **action** of the mechanical system along the curve α.

Now, the principle

Hamilton's Principle Of Least Action Among all the curves in $C([a, b])$ the mechanical system will move along that curve along which its action is least.

We use this principle to obtain Lagrange's equations of motion as follows.

Suppose, $\alpha : [a,b] \longrightarrow C$ is the trajectory of the system. We consider any smooth map $\eta : [a,b] \longrightarrow \mathbb{R}^N$, $\eta(t) = \left(\eta^1(t), \eta^2(t) \cdots \eta^N(t)\right)$ satisfying $\eta(a) = \eta(b) = 0$. *Then for all $s \in \mathbb{R}$, with $|s|$ sufficiently small, we get $\alpha(t) + s\eta(t) \in C$ for all t in $[a,b]$.*

Note that $C \subset \mathbb{R}^N$ and so, $\alpha(t) \in C(\subset \mathbb{R}^N)$, $s\eta(t) \in R^N$ and so, $\alpha(t) + s\eta(t)$ is a well-defined element of $\mathbb{R}^N$. In fact, compactness of $[a,b]$ and openers of C in $\mathbb{R}^N$ provides us with a constant $\delta > 0$ such that $\alpha(t) + s\eta(t) \in C$ for all $t \in [a,b]$ and for all $|s| < \delta$.

Thus, we get a family $\{\alpha_s : [a,b] \longrightarrow C : |s| < \delta\}$; all α_s being smooth curves in C joining A to B. Now the action of the mechanical system along each α_s determines the map

$$(-\delta, \delta) \longrightarrow \mathbb{R}; s \longmapsto A(\alpha_s)$$

According to Hamilton's principle of least action, we have $\frac{d}{ds}A(\alpha_s)_{s=o} = 0$. We calculate $\frac{d}{ds}A(\alpha_s)$ first

$$
\begin{aligned}
\frac{d}{ds}\left(A(\alpha_s)\right) &= \frac{d}{ds}\int_a^b L\left(\alpha_s(t), \dot{\alpha}_s(t)\right) dt \\
&= \int_a^b \frac{\partial}{\partial s} L\left(\alpha_s(t), \dot{\alpha}_s(t)\right) dt \\
&= \int_a^b \left\{ \sum_{i=1}^N \frac{\partial L}{\partial q^i} \frac{\partial}{\partial s}\alpha_s^i(t) + \sum_{i=1}^N \frac{\partial L}{\partial \dot{q}^i} \frac{\partial}{\partial s}\dot{\alpha}_s^i(t) \right\} dt \\
&= \sum_{i=1}^N \int_a^b \frac{\partial L}{\partial q^i} \frac{\partial a l_s^i(t)}{\partial s} dt + \sum_{i=1}^N \frac{\partial L}{\partial \dot{q}^i} \frac{\partial}{\partial s} \frac{\partial}{\partial t}\alpha_s^i(t)dt \\
&= \sum_{i=1}^N \int_a^b \frac{\partial L}{\partial q^i} \frac{\partial a l_s^i(t)}{\partial s} dt + \sum_{i=1}^N \int_a^b \frac{\partial L}{\partial \dot{q}^i} \frac{\partial}{\partial t} \frac{\partial}{\partial s}\alpha_s^i(t)dt \\
&= \sum_{i=1}^N \int_a^b \frac{\partial L}{\partial q^i} \frac{\partial a l_s^i(t)}{\partial s} dt + \sum_{i=1}^N \left[\frac{\partial L}{\partial \dot{q}^i} \frac{\partial \alpha_s^i(t)}{\partial s}\right]_a^b \\
&\quad - \sum_{i=1}^N \int_a^b \left(\frac{\partial}{\partial t} \frac{\partial L}{\partial \dot{q}^i}\right) \frac{\partial \alpha_s^i(t)}{\partial s} dt
\end{aligned}
$$

using integration by parts

$$= \sum_{i=1}^N \int_a^b \left[\frac{\partial L}{\partial q^i} - \frac{\partial}{\partial t}\frac{\partial L}{\partial \dot{q}^i}\right] \frac{\partial \alpha_s^i(t)}{\partial s} dt + 0$$

since $\alpha_s(a) = A, \alpha_s(b) = B$ for all $s \in (-\delta, \delta)$ implies $\frac{\partial \alpha^i}{\partial s}(a) = 0 = \frac{\partial \alpha^i}{\partial s}(b)$.

Now $\alpha_s(t) = \alpha(t) + s\eta(t)$ implies $\alpha_s^i(t) = \alpha^i(t) + s\eta^i(t)$ and hence

$\frac{\partial \alpha_s^i(t)}{\partial s} = \eta^i(t)$. Therefore

$$\frac{d}{ds} A(\alpha_s) = \sum_{i=1}^{N} \int_a^b \left[\frac{\partial L}{\partial q^i} - \frac{\partial}{\partial t} \frac{\partial L}{\partial \dot{q}^i} \right] \eta^i(t) = 0.$$

The equation

$$\sum_{i=1}^{N} \int_a^b \left\{ \frac{\partial L}{\partial q^i}(\alpha(t), \dot{\alpha}(t)) - \frac{d}{dt} \frac{\partial L}{\partial \dot{q}^i}(\alpha(t), \dot{\alpha}(t)) \right\} \eta^i(t) dt = 0$$

is valid for all $\eta : [a, b] \longrightarrow \mathbb{R}^N$ satisfying $\eta(a) = \eta(b)$. In particular, for any $k(1 \leq k \leq N)$, taking $\eta(t) = (0, \dots, 0, \eta^k(t), 0 \dots 0)$ with $\eta^k : [a, b] \longrightarrow \mathbb{R}$ arbitrary, except that it satisfies $\eta^k(a) = \eta^k(b) = 0$, the sum on the left hand side of the last equation reduces to the single, (k^{th}) summand and the equation reduces to

$$\int_a^b \left\{ \frac{\partial L}{\partial q^k}(\alpha(t), \dot{\alpha}(t)) - \frac{d}{dt} \frac{\partial L}{\partial \dot{q}^k}(\alpha(t), \dot{\alpha}(t)) \right\} \eta^k(t) d_t = 0.$$

Again, this being true for all $\eta^k : [a, b] \longrightarrow \mathbb{R}$ with $\eta^k(a) = \eta^k(b) = 0$, we get

$$\frac{\partial L}{\partial q^k}(\alpha(t), \dot{\alpha}(t) - \frac{d}{dt} \frac{\partial L}{\partial \dot{q}^k}(\alpha(t), \dot{\alpha}(t)) = 0, 1 \leq k \leq N$$

thus proving that the trajectory $t \longmapsto \alpha(t)$ of the system satisfies Lagrange's equations (14).

We have now explained how Hamilton's principle gives rise to Lagranges equations of motion.

7.7   Hamilton's Equations of Motion

Lagranges equations form a set of second order differential equations. We want to associate with them an equivalent set of first order differential equations, the number of the latter equations being double the former. We explain the procedure below:

Definition 4 The quantity $\frac{\partial L}{\partial \dot{q}^i}$ is called the **generalized momentum** conjugate to the generalized coordinate q^i.

We denote it by p_i. Thus, $p_i = \frac{\partial L}{\partial \dot{q}^i}$. Note that if a generalized coordinate is a Cartesian coordinate say x^k, then the generalized momentum conjugate to x^k is $m_k \dot{x}^k$, the x-component of the linear momentum of the particle P_k. Thus the concept of generalized momentum is indeed a generalization of the Newtonian concept of momentum of a particle.

Let M denote the set of all ordered $2N$-tuples $(q, \dot{q}) = (q^1, q^2, \cdots q^N, \dot{q}^1, \dot{q}^2 \cdots \dot{q}^N)$ consisting of a position q and a velocity $\dot{q}$ of the system while passing through q.

Definition 5 M is the **velocity phase space** of the mechanical system. Also, let M^* be the set of all the ordered $2N$-tuples $(q,p) = (q^1, q^2, \cdots, q^N, p^1, p^2, \cdots p^N)$ where p is the generalized momentum (that is the N-tuple of generalized momenta $(p^1, p^2, \cdots, p^N)$) of the system while it is at the position q.

Definition 6 M^* is called the **momentum phase space** of the mechanical system.

An element $m \in M$ represents the instantaneous state of motion of the mechanical system in terms of its instantaneous position q and the associated (generalized) velocity $\dot{q}$. Similarly, an element η of M^* represents the instantaneous state of motion of the system in terms of the instantaneous position together with the associated generalized momentum.

Now we use the Lagrangian function $L(q, \dot{q})$ of the system to define a map

$$\mathcal{L} : M \longrightarrow M^*$$

which is given by $\mathcal{L}(q, \dot{q}) = (q, p)$.

The function $\mathcal{L}$ is called the **Legendre transformation** of the mechanical system.

We will prove below that $\mathcal{L}$ is a bijective map. Towards this goal, we consider the Lagrangian

$$\frac{1}{2} \sum_{i,j=1}^{N} a_{ij}(q) \dot{q}^i \dot{q}^j - U(q^1 \cdots q^N).$$

Now $p_i = \frac{\partial L}{\partial \dot{q}^i} = \sum_{j=1}^{N} a_{ij}(q) \dot{q}^j$, $1 \le i \le N$. Using the matrix $A(q) = [a_{ij}(q)]$; we get $p = A(q)\dot{q}$. But being symmetric and positive definite, the matrix $A(q)$ is invertible and consequently we have both

$$p = A(q)\dot{q} \text{ and } \dot{q} = A(q)^{-1}p \qquad (18)$$

Having noticed the above relations, we prove the injectivity property of the Legendre transformation $\mathcal{L}$: Let $m = (q, \dot{q})$ and $\tilde{m} = (\tilde{q}\dot{\tilde{q}})$ be any two elements of M. Then $\mathcal{L}(m) = (q, A(q)\dot{q})$ and $\mathcal{L}(\tilde{m}) = (\tilde{q}, \alpha(\tilde{q})\dot{\tilde{q}})$. Hence $\mathcal{L}(m) = \mathcal{L}(\tilde{m})$ will imply $q = \tilde{q}$, $A(q)\dot{q} = A(\tilde{q})\dot{\tilde{q}} = A(q)\dot{\tilde{q}}$ leading to $q = \tilde{q}$ and $\dot{q} = \dot{\tilde{q}}$. This proves the injective property of $\mathcal{L}$.

Finally if $\eta = (q, p)$ is an element of M^*, then we consider $m = (q, A(q)^{-1}p) \in M$. We then have $\mathcal{L}(m) = \eta$. Hence $\mathcal{L}$ is surjective (onto) also.

Now we have proved the result that the map $\mathcal{L} : M \longrightarrow M^*$ is a bijection. In passing, let us note that relations (18) express two simple facts namely, the generalized momenta are functions of generalized coordinates and the generalized momenta that is $p_i = p_i(q, \dot{q})$ and in turn, the generalized velocities $\dot{q}^i$ are functions of the generalized coordinates and momenta that is, $\dot{q}^i = \dot{q}^i(q, p)$.

Now we define a function $H : M^* \longrightarrow \mathbb{R}$ by

$$H(q,p) = \sum_{i=1}^{N} \dot{q}^i p^i - L(q,p).$$

Here, the $\dot{q}^i$ appearing on the right hand side of the relation defining H are to be treated as functions of (q,p) and so, H is indeed a function defined on the momentum phase space M^* of the system.

Definition 7 The function H is called **Hamiltonian function** of the mechanical system.

We consider the differential of H which we obtain in two different ways. First, using the dependence of H on its variables $(q^1, q^2, \cdots, q^N, p^1, p_2, \cdots p_N)$, we get

$$dH = \sum_{i=1}^{N} \frac{\partial H}{\partial q^i} dq^i + \sum_{i=1}^{N} \frac{\partial H}{\partial p^i} dp_i \qquad (19)$$

On the other hand using the definition of $H = \sum_{i=1}^{n} q^i p_i - L$, we get

$$dH = \sum_{i=1}^{N} p_i d\dot{q}^i + \sum_{i=1}^{N} \dot{q}^i dp_i - \sum_{i=1}^{N} \frac{\partial L}{\partial \dot{q}^i} d\dot{q}^i.$$

Now the system satisfies Lagranges equations and so, we can substitute $\dot{p}_i$ in place of $\frac{\partial L}{\partial q^i}$ in the third summation of the above differential equation. Also, we substitute p_i in place of $\frac{\partial L}{\partial \dot{q}^i}$ in the fourth summation

$$dH = \sum_{i=1}^{N} p_i d\dot{q}^i + \sum_{i=1}^{N} \dot{q}^i dp_i - \sum_{i=1}^{N} \dot{p}_i dq^i - \sum_{i=1}^{N} p_i d\dot{q}^i$$

and therefore, we get

$$dH = \sum_{i=1}^{N} \dot{q}^i dp_i - \sum_{i=1}^{N} \dot{p}_i dq^i \qquad (20)$$

Combining (19) and (20) we get

$$\sum_{i=1}^{N} \frac{\partial H}{\partial q^i} dq^i + \sum_{i=1}^{N} \frac{\partial H}{\partial p_i} dp_i = \sum_{i=1}^{N} \dot{q}_i dp_i - \sum_{i=1}^{N} p_i dq^i.$$

Comparing the coefficients of dq^i and dp_i on both sides of the above equation, we get:

$$\dot{q}^i = \frac{\partial H}{\partial p_i}, \dot{p}_i = -\frac{\partial H}{\partial q^i} \quad 1 \le i \le N.$$

The above equations are the desired first order ordinary differential equations in the position coordinates $(q^1, q^2, \cdots q^N)$ and the associated generalized momenta $(p_1, p_2, \cdots p_N)$. They are called **Hamilton's equations of motion.**

We deduced Hamilton's equations of motion from Lagrange's equations. Conversely, it can be shown that Hamilton's equations give rise to Lagrange's equations of motion. The deduction of Lagrange's equations from Hamilton's equations is left as the Exercise 11 at the end of this chapter.

Thus, there are two different but equivalent ways of obtaining information about the motion

(1) By solving the second order ordinary differential equations, that is, the Lagrange's equations.

(2) Solving the (first order but twice as many) Hamilton's equations.

Example 4 A particle P of mass m is constrained to move on the cylinder $x^2 + y^2 = a^2$. It is acted upon by a force which is directed towards a fixed point O on the axis of the cylinder; the magnitude of the force being proportional to the distance OP of the particle P from O. Obtain Hamiltons equations of motion, and solving the equations, describe the motion.

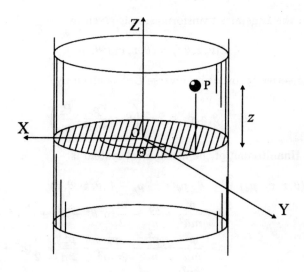

Figure 7.8

Solution We take the pair (z, θ) as the generalized coordinates. They are as shown in Fig. 7.8. Writing $\vec{r}$ for $\vec{OP}$, we have the form of the force field.

$$F(\vec{r}) = -\alpha \vec{r}.$$

α being the positive constant of proportionality. Clearly, the force field is conservative. In fact, a potential energy function U for the force field is given by

$$U(\vec{r}) = \frac{\alpha}{2}(x^2 + y^2 + z^2) = \frac{\alpha}{2}(a^2 + z^2), \quad \vec{r} = x\vec{i} + y\vec{j} + z\vec{k}$$

Note that $x = a\cos\theta, y = a\sin\theta$ and consequently $\dot{x} = -a\dot{\theta}\sin\theta$ and $y = a\dot{\theta}\cos\theta$. Now, the kinetic energy T of the particle is given by

$$
\begin{aligned}
T &= \frac{m}{2}\left[\dot{x}^2 + \dot{y}^2 + \dot{z}^2\right] = \frac{m}{2}\left[a^2\dot{\theta}^2\sin^2\theta + a^2\dot{\theta}^2\cos^2\theta + \dot{z}^2\right] \\
&= \frac{m}{2}\left[a^2\dot{\theta}^2 + \dot{z}^2\right].
\end{aligned}
$$

The Lagrangian function is

$$L(\theta, z, \dot{\theta}, \dot{z}) = \frac{m}{2}\left[a^2\dot{\theta}^2 + \dot{z}^2\right] - \frac{\alpha}{2}\left[a^2 + z^2\right].$$

Let p_θ be the generalized momentum conjugate to θ and let p_z be that conjugate to z. They are given by

$$p_\theta = \frac{\partial L}{\partial \dot{\theta}} = ma^2\dot{\theta} \text{ and } p_z = \frac{\partial L}{\partial \dot{z}}.$$

Therefore, the Legendre transformation is given by

$$\mathcal{L}(\theta, z, \dot{\theta}\dot{z}) = (\theta, z, ma^2\dot{\theta}, m\dot{z}).$$

$\mathcal{L}$ being bijective (proved) its inverse $\mathcal{L}^{-1}$ is given by

$$\mathcal{L}^{-1}(\theta, z, p_\theta, p_z) = \left(\theta, z, \frac{p_\theta}{ma^2}, \frac{P_z}{m}\right).$$

Now, the Hamiltonian of the mechanical system is

$$
\begin{aligned}
H(\theta, z, p_\theta, p_z) &= \dot{\theta} \cdot p_\theta + \dot{z} \cdot p_z - L(\theta, z\cdot, \dot{\theta}, \dot{z}) \\
&= \frac{p^2}{ma^2} + \frac{p_z^2}{m} - \frac{m}{2}(a^2\dot{\theta}^2 + \dot{z}^2) + \frac{\alpha}{2}(a^2 + z^2) \\
&= \frac{p^2}{ma^2} + \frac{p_z^2}{m} - \frac{p_\theta^2}{2ma^2} - \frac{p_z^2}{2m} + \frac{\alpha}{2}(a^2 + z^2) \\
&= \frac{p_\theta^2}{2ma^2} + \frac{p_z^2}{2m} + \frac{\alpha}{2}(a^2 + z^2).
\end{aligned}
$$

Thus, $H(\theta, z, p_\theta, p_z) = \frac{p_\theta^2}{2ma^2} + \frac{p_z^2}{2m} + \frac{\alpha}{2}(a^2 + z^2)$. Hamiltons equations are

(i) $\dot{\theta} = \frac{\partial H}{\partial p_\theta}$, that is $\dot{\theta} = \frac{p_\theta}{ma^2} = \frac{ma^2\dot{\theta}}{ma^2} = \dot{\theta}$

(ii) $\dot{z} = \frac{\partial H}{\partial p_z}$, that is $\dot{z} = \frac{p_z}{m} = \dot{z}$.

[Thus the first two Hamilton's equations do not give any information in this example.]

(iii) $\dot{p}_\theta = -\frac{\partial H}{\partial \theta} = 0$ since H is independent of θ. Therefore p_θ ($= p_\theta(t)$) remains constant say equal to β. Thus, $ma^2\dot{\theta} = \beta$ which gives $\dot{\theta} = \frac{\beta}{ma^2}$. Hence we get that the particle moves on the cylindrical surface in such a way that its rotational motion around the axis of the cylinder is with constant angular speed.

(iv) $\dot{p}_z = -\frac{\partial H}{\partial z}$ gives $m\ddot{z} = -\alpha z$ which has the solution $z(t) = R\cos\left(\sqrt{\left(\frac{\alpha}{m}\right)} \cdot t + \gamma\right)$, where R and γ are some constants.

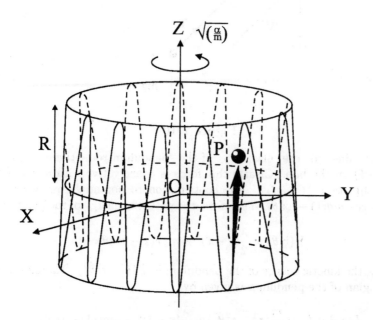

Figure 7.9

The solution $z(t) = R\cos\left(\sqrt{\left(\frac{\alpha}{m}\right)} \cdot t + \gamma\right)$ implies that the z-component of the motion is simple harmonic in nature, R being its amplitude and $\frac{\left(\sqrt{\frac{\alpha}{m}}\right)}{2\pi}$ its frequency. Thus, the motion of the particle is a combination of (a) circular motion about the axis of the cylinder, the angular velocity of the circular motion being constant and (b) a simple harmonic motion parallel to the axis of the cylinder.

Example 5 A pendulum consists of an elastic rod having natural length ℓ and modulus of elasticity λ; its bob having mass m (while the mass of the rod is negligible). The pendulum is hanged from a fixed point O and is free to oscillate about O in a fixed vertical plane (through O). Obtain

(i) an expression for the Lagrangian function for the pendulum and (ii) the associated Lagrange's equations of motion.

Solution Let $r = r(t)$ be the (extended) instantaneous length of the pendulum and let $\theta(t)$ be the angle subtended by the rod with the vertical. Then we can take the pair (r, θ) as a set of generalized coordinates.

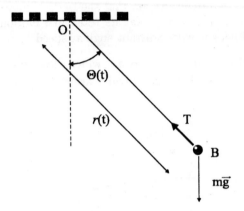

Figure 7.10

Now, due to the elasticity of the pendulum rod, there is a force $-\lambda(r - \ell)$ on the bob. Clearly this force is conservative with a potential energy function $\frac{\lambda}{2}(r - \ell)^2$. The contribution to potential energy due to gravity is $-mgr(1 - \cos\theta)$. Therefore, the potential energy of the pendulum is

$$Y(r, \theta) = \frac{\lambda}{2}(r - \ell)^2 - mgr(1 - \cos\theta).$$

Clearly, the kinetic energy of the pendulum is $\frac{m}{2}(\dot{r}^2 + r^2\dot{\theta}^2)$. Therefore, the Lagrangian of the pendulum is given by

$$L(r, \theta, \dot{r}, \dot{\theta}) = \frac{m}{2}(\dot{r}^2 + r^2\dot{\theta}^2) - \frac{\lambda}{2}(r - \ell)^2 + mgr(1 - \cos\theta).$$

The Lagranges equations are

$$\frac{d}{dt}\left(\frac{\partial L}{\partial \dot{r}}\right) - \frac{\partial L}{\partial r} = 0 \text{ and } \frac{d}{dt}\left(\frac{\partial L}{\partial \dot{\theta}}\right) - \frac{\partial L}{\partial \theta} = 0$$

which take the form:

$$\begin{aligned}
m\ddot{r} - mr\dot{\theta}^2 + \lambda(r - \ell) - mg(1 - \cos\theta) &= 0 \\
mr^2\ddot{\theta} - mgr\sin\theta &= 0 \quad \text{or equivalently} \\
r\ddot{\theta} - g\sin\theta &= 0.
\end{aligned}$$

□

Example 6 A particle P of mass m is sliding on the smooth surface of revolution of the parabola $x^2 + y^2 = az$ with its axis vertical. The particle is moving under the gravitational force. Taking the polar coordinates (r, θ) as the generalized coordinates, obtain the Lagrangian equations of motion.

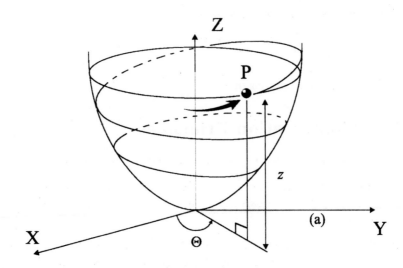

Z

P

z

X

Θ

Y

(a)

Figure 7.11

Solution The potential energy of the particle is $U = mgz = mg\frac{r^2}{a}$. The kinetic energy is given by

$$T = \frac{m}{2}\left[\dot{x}^2 + \dot{y}^2 + \dot{z}^2\right] = \frac{m}{2}\left[\dot{r}^2 + r^2\dot{\theta}^2 + \dot{z}^2\right]$$

$$= \frac{m}{2.}\left[\dot{r}^2 + r^2\dot{\theta}^2 + \frac{4r^2}{a^2}\cdot\dot{r}^2\right] = \frac{m}{2}\left[\left(1 + \frac{4r^2}{a^2}\right)\dot{r}^2 + r^2\dot{\theta}^2\right].$$

Therefore, the Lagrangian function of the particle is

$$L(r, \theta, \dot{r}, \dot{\theta}) = \frac{m}{2}\left[\left(1 + \frac{4r^2}{a^2}\right)\dot{r}^2 + r^2\dot{\theta}^2\right] - mg\frac{r^2}{a}.$$

The two Lagrange's equations are

$$\frac{d}{dt}\left(\frac{\partial L}{\partial \dot{r}}\right) - \frac{\partial L}{\partial r} = 0 \text{ and } \frac{d}{dt}\left(\frac{\partial L}{\partial \dot{\theta}}\right) - \frac{\partial L}{\partial \theta} = 0. \qquad (*)$$

The first of them takes the form

$$m\left(1 + 4\frac{r^2}{a^2}\right)\ddot{r} + \frac{4mr\dot{r}^2}{a^2} - mr\dot{\theta}^2 + mg\frac{r}{a} = 0$$

or equivalently

$$(a^2 + 4r^2)\ddot{r} + 4\dot{r}^2 - a^2r\dot{\theta}^2 + gar = 0. \qquad (**)$$

Also, $\frac{\partial L}{\partial \dot{\theta}} = mr^2\dot{\theta}$ while, $\frac{\partial L}{\partial \theta} = 0$. Therefore, the second equation in
(*) gives $\frac{d}{dt}(mr^2\dot{\theta}) = 0$ or $mr^2\dot{\theta} = $ constant. Hence the second Lagrange's
equation becomes

$$r^2\dot{\theta} = h \qquad\qquad (***)$$

Combining (**) and (***), we get

$$(a^2 + 4r^2)\ddot{r} + 4r\dot{r}^2 - \frac{a^2h^2}{r^3} + gar = 0 \text{ with } r^2\dot{\theta} = h.$$

These are the equations of motion.   □

7.8   Conservation Principles

With the particle system (1), we have associated two functions.

(1) The Lagrangian function $L : M \longrightarrow \mathbb{R}$ defined on the velocity
    phase spare M consisting of all $2N$-tuples $(q, \dot{q}) = (q^1, q^2, \cdots, q^N,$
    $\dot{q}^1, \dot{q}^2, \cdots, \dot{q}^N)$.

(2) The Hamiltonian function $H : M^* \longrightarrow \mathbb{R}$ defined on the momentum
    phase space M^* consisting of the $2N$-tuples $(q, p) = (q^1, q^2, \cdots, q^N,$
    $p^1, p^2, \cdots, p^N)$.

Recalling the equation relating L and H, namely,

$$H(q, p) = \sum_{i=1}^{N} p_i \dot{q}^i - L(q, \dot{q}),$$

we see at once that for any $j(1 \leq j \leq N)$ if the Lagrangian function is
independent of q^j, then so is the Hamiltonian function. The converse also
holds. Thus $\frac{\partial L}{\partial q^j} \equiv 0$ if and only if $\frac{\partial H}{\partial q^j} \equiv 0$.

Definition 8 A coordinate q^j is said to be **cyclic** if $\frac{\partial L}{\partial q^j} \equiv 0$ (and hence
$\frac{\partial H}{\partial q^j} \equiv 0$ also).

Now, the Lagraange's equation $\frac{d}{dt}\frac{\partial L}{\partial \dot{q}^j} = \frac{\partial L}{\partial q^j}$ (or equivalently, the Hamilton's equation $\dot{p}_j = -\frac{\partial H}{\partial q^j}$) implies that the observable $p_j : M \longrightarrow \mathbb{R}$ is a
first integral of motion. (in M^* the *coordinate* function $p_j : M^* \longrightarrow \mathbb{R}$
remains constant along each trajectory of the system).

Thus, we have proved the following proposition:

Proposition 1 The momentum conjugate to a cyclic coordinate is a first
integral of motion.

Recall, in **Example 6** in this chapter, the angular coordinate θ is cyclic
and consequently the momentum conjugate to it, namely $mr^2\dot{\theta}$ is a constant
h. We used the constancy of it to eliminate $\dot{\theta}$ from the equation (**) and

simplified to the ODE $(* * *)$. Recall also, that we employed a similar technique in the theory of central orbits. In fact, in a plane, we chose the polar coordinates (r, θ) and we had the Lagrangian

$$L(r, \theta, \dot{r}, \dot{\theta}) = \frac{m}{2} \left[\dot{r}^2 + r^2 \dot{\theta}^2 \right] + \psi(r).$$

Notice that θ is a cyclic coordinate, giving rise to the first integral $L = mr^2 \dot{\theta} =$ momentum conjugate to the generalized coordinate θ which was used to simplify the equations of motion.

Thus, cyclic coordinates give rise to first integrals of motion and the first integrals in turn simplify the equations governing the motion. In passing, we prove the following result now:

Proposition 2 $H : M^* \longrightarrow \mathbb{R}$ is a first integral of motion.

Proof We consider a $t \longmapsto (q(t), p(t))$ of the system in the momentum phase space. Differentiating H along this trajectory; we get

$$\frac{d}{dt} H(q(t), p(t)) = \sum_{i=1}^{N} \frac{\partial H}{\partial q^i} \frac{dq^i}{dt} + \sum_{i=1}^{N} \frac{\partial H}{\partial p_i} \frac{dp_i}{dt}$$

$$= \sum_{i=1}^{N} \frac{\partial H}{\partial q^i} \frac{\partial H}{\partial p_i} + \sum_{i=1}^{N} \frac{\partial H}{\partial p_i} \left(-\frac{\partial H}{\partial q^i} \right)$$

using Hamiltonian's equations

$$\equiv 0.$$

Thus, along any trajectory $t \longmapsto q(t), p(t)$, the function H is constant, proving the result. ☐

Recall that $p_i = \frac{\partial L}{\partial \dot{q}^i} = \frac{\partial}{\partial \dot{q}^i} \left\{ \frac{1}{2} \sum_{1,j=1}^{N} a_{ij}(q) \dot{q}^i \dot{q}^j - U(q) \right\} = \sum_{j=1}^{N} a_{ij}(q) \dot{q}^j$

and hence $\sum_{i=1}^{N} p_i \dot{q}^i = \sum_{i,j=1}^{N} a_{ij}(a) \dot{q}^i \dot{q}^j = 2T$ (the kinetic energy of the system consequently,

$$H = \sum_{i=1}^{N} p_i \dot{q}^i - L = 2T - (T - U) = T + U = E.$$

Thus we have verified the following: The Hamiltonian H which is a function of (q, p) when expressed in term of $(q, \dot{q})$ becomes the total energy of the mechanical system. In other words $H \circ \mathcal{L} = E$.

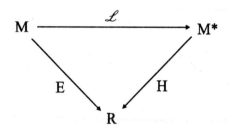

Concluding Remark In this chapter we discussed two types of differential equations governing the motion: Lagrange's equations which are second order ordinary differential equations on the configuration space C of the mechanical system and Hamilton's equations which are defined on the (entirely conceptual) momentum phase space M^*. In both the cases, the domains of the differential equations (that is, C and M^*) are spaces with intricate geometric properties. Consequently, along with the physical agencies (that is, the forces, the energy, etc.) the geometry of these spaces also influences the dynamics of the system. As a consequence, the differential equations are more difficult to solve and get the information about the motion. To proceed further, we need know a lot more about the modern treatment of differential equations. In fact, the theory of differential equations is so intimately related with the dynamics of the entire physical world that in modern nomenclature, ordinary differential equations are better known as smooth **dynamical systems**. May we conclude this chapter by quoting V.I. Arnold (an eminent Russian mathematician)

"Newton's fundamental discovery , the one which he considered necessary to keep secret and published only in the form of an anagram consists of the following : *Data aequatione quotinque fluentes quantitae involvente fluxiones invenire at vice versa*".

In contemporary mathematica language, this means

" **It Is Useful To Solve Differential Equations.**"

EXERCISES

1. A satellite of mass m is orbiting the earth, the distance of the satellite from the center of the earth being denoted by r. Let the mass of the earth be M so that the force of gravity of the earth acting on the satellite is $-\gamma \frac{Mm}{r^2}$ along the line towards the center of earth γ being the usual gravitational constant. Suppose the satellite orbits the earth in a plane through the center of the earth. Let (r, θ) be the polar coordinates about the center (in the plane of motion). Show that the Lagrangian of the satellite can be written as

$$L(r, \theta, \dot{r}, \dot{\theta}) = \frac{m}{2}(\dot{r}^2 + r^2\dot{\theta}^2) + \frac{Mm}{r}.$$

Obtain Lagrange's equations of motion for the satellite and deduce that $r^2\dot\theta$ is a constant quantity.

2. Obtain the Lagrangian and Hamiltonian functions for a particle of mass m swinging under gravity when attached to a fixed point O by a light straight string of variable length $\ell(t)$ where its motion is confined to a vertical plane through O.

3. Find the Hamiltonian and Hamilton's equations of motion of a partical of mass m sliding under gravity on a smooth rigid parabolic wire in the shape $z = \frac{1}{2}\alpha^2 x^2$ in the vertical XOZ-plane, the Z-axis being vertically upwards.

4. A particle of mass m moves under gravity along the smooth spiral wire, its shape being defined by the angular parameter θ by $x = a \cdot \cos\theta, y = a\sin\theta, z = k$, a and k being positive constants. Obtain Hamilton's equations of motion.

5. A particle of mass m is constrained to move in the vertical XOY-plane(Y-axis vertically upwards under the influence of gravity along a given curve. The curve has parametric equations: $x = f(\eta), y = h(\eta)$. Taking η as the generalized coordinate, show that the Lagrangian of the particle is given by

$$L = \frac{1}{2}m\dot\eta^2 \left[\left(\frac{d\rho}{d\eta}\right)^2\right] - mgh(\eta).$$

Find the corresponding Hamiltonian.

Obtain Lagrange's and Hamilton's equations of motion.

6. Two configuration coordinates q and Q of the same mechanical system (having one degree of freedom) are related by the equation $q = f(Q)$. If $L(q, \dot q)$ and $\vec L(Q, \dot Q)$ are the Lagrangian functions of the same system expressed in the coordinates q and Q respectively (therefore $\vec L(Q, \dot Q) = L(f(Q), f'(Q)\dot Q)$, deduce the equation $\frac{d}{dL}\left(\frac{\partial \vec L}{\partial \dot Q}\right) - \frac{\partial L}{\partial Q} = 0$ from the equation $\frac{d}{dt}\left(\frac{\partial L}{\partial \dot q}\right) - \frac{\partial L}{\partial q} = 0$. Also show that the momenta p and P conjugate to q and Q respectively are related by the equation $p = \frac{P}{f(Q)}$.

7. If a mass m is projected vertically upwards, prove that the Lagrangian of the particle is given by $L = \frac{m}{2}\dot z^2 + \frac{\gamma M m}{(R+z)}$ γ being the gravitational constant, M and R, the mass and the radius of the earth and z, the height of the mass above the surface of the earth. Find an expression for the Hamiltonian function and also Hamilton's equations of motion.

8. The Lagrangian of a particle is $L = \frac{1}{2}\dot{q}^2 + \frac{1}{2}q^3$. It starts at $t = 0$ from the point $q = 1$ with the generalized momentum $p = 1$. Obtain that the motion terminates when $t = T$ and find T.

9. Find the Lagrangian and Hamiltonian of a pendulum comprising a mass m attached to a light stiff rod AB having length ℓ free to move in a vertical plane in such a way that the end A of the rod is forced to move vertically, its distance from a fixed point O being a given function $\gamma(t)$ of time t. Taking the angle θ as the generalized coordinate (see fig.7.12), obtain the Lagrangian and Hamiltonian equations of motion.

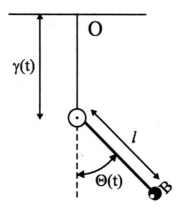

Figure 7.12

10. The Lagrangian function of a particle is given by $L = \frac{m}{2}\dot{q}^2 - Aq^4$, $A > 0$ (one degree of freedom).

    The particle starts at time $t = 0$ from a point $q \cdot > 0$ with momentum $p > 0$. Obtain an expression for the function $t \longmapsto q(t)$.

11. Deduce Lagrange's equations from Hamilton's equation.

Supplementary Reading

Here is a short list of books for supplementary reading [2] gives interesting historical development of the subject [1] is a graduate level text-book. It is included here because, we feel that an interested reader can get some ideas about further developement of the subject from such a book. [3], [5] are rather old fashioned books, but they also give another view-point of the subject.

[1] V.I. Arnold, Mathematical Methods of Classical Mechanics.

[2] Dugals. R.A., History of Mechanics.

[3] Loney S.L., Dynamics.

[4] Lunn Mary, A First Course in Mechanics.

[5] Ramsey S.L., Dynamics.

[6] Synge J.L. and Griffiths B.A., Principles of Mechanics.

[7] Woodhouse N.M.J., Introduction to Analytical Dynamics.

Index